The Bohr Atom

A guide

The Bohr Atom

A guide

Bruce Cameron Reed

Department of Physics (Emeritus), Alma College, Alma, MI, USA

IOP Publishing, Bristol, UK

ISBN 978-0-7503-3611-6 (ebook)
ISBN 978-0-7503-3609-3 (print)
ISBN 978-0-7503-3612-3 (myPrint)
ISBN 978-0-7503-3610-9 (mobi)

DOI 10.1088/978-0-7503-3611-6

Version: 20201101

IOP ebooks

British Library Cataloguing-in-Publication Data: A catalogue record for this book is available from the British Library.

Published by IOP Publishing, wholly owned by The Institute of Physics, London

IOP Publishing, Temple Circus, Temple Way, Bristol, BS1 6HG, UK

US Office: IOP Publishing, Inc., 190 North Independence Mall West, Suite 601, Philadelphia, PA 19106, USA

This work is dedicated to Laurie. Again.

Contents

Preface

Every student of physics or chemistry encounters the Bohr model of the hydrogen atom: a lone electron orbits a nucleus containing one or more protons, restricted to orbits whose radii are 'quantized' in that they are restricted to particular values. Energy in the form of particle-like 'photons' is emitted to the outside world as the electron 'falls' from a higher-radius orbit to a lower-radius one, or conversely absorbs energy from an outside source of light and 'jumps' to a higher-radius orbit. In more advanced courses, students learn that Bohr's model was eventually superseded by the probabilistic particle-wave 'quantum mechanics' of Heisenberg, Schrödinger, Born, Jordan, and others. But Bohr's model was the first mathematical treatment of atomic structure which made accurate predictions, albeit at the cost of introducing revolutionary ideas into physics. For these reasons alone, the place of the Bohr model in the canon of physics is well-deserved. Beyond this, the Bohr model stands as a superb example of how scientific theories develop. In the beginning are well-established empirical facts; in this case regularities in spectral features and the chemical behavior of elements. A provisional theoretical model incorporating elements of both established physics and some new *ad hoc* assumptions is proposed, and proves surprisingly successful at 'explaining' some but by no means all of the empirical facts. Tests of the theory lead to refinements, which eventually culminate in the provisional theory being succeeded by a more complete one which often retains technical and/or conceptual elements of the original version.

An online search on 'Bohr atomic model' will return millions of hits, and the topic is covered in libraries of textbooks. Why did I feel that it deserves yet another treatment? Fundamentally, there are two reasons.

First: for many years, I taught the Bohr atom in modern physics, quantum mechanics, and astronomy classes at various levels of undergraduate physics curricula. Curiosity drove me to read Bohr's original papers, and I came away from that effort with a sense that modern texts give a very narrow and incomplete picture as to what motivated Bohr and what he actually did. For instance, derivations of the Bohr atom often begin by assuming quantization of the electron's orbital angular momentum, an argument which, when combined with classical centripetal force and energy arguments, leads to the 'correct' quantized orbital radii and transition energies. However, this is not at all what Bohr did. Rather, his original derivation involved two steps: a curious 'binding-of-the-electron-to-the-nucleus' argument, and then consideration of orbital transitions. Both steps made use of Max Planck's quantum-of-radiation expression $E = hf$ (more on this in section 1.2), but in different ways. These comments are not intended to be critical of texts. Such works are carefully prepared by teams of experienced, knowledgeable authors and editors. Many scientific breakthroughs begin life convoluted and muddled, and authors have to clean things up: it would be impossible in either a text or class to work through every wrinkle of how a physical theory developed. When I was a student, I recall one of my professors remarking that discoveries are like diamonds: when they come out of the ground, they are covered in mud and full of rough edges; what ultimately makes it into texts are the cleaned and polished versions. Detailed reconstructions of the developments of theories and experiments are usually the province of specialists and professional historians of science who produce lengthy books and journal papers in order to impress tenure committees. Since most students and working scientists do not have the time to dig into the history of some area, I felt it would be valuable to develop a treatment of the Bohr atom that is more detailed than what is encountered in standard texts, while maintaining a level of brevity and mathematical accessibility that would make the presentation suitable as supplementary material for students and instructors.

Second: Bohr lived for nearly 50 years after the publication of his hydrogenic model, and enjoyed a remarkably rich life and career. I felt it equally appropriate to offer readers a fuller if necessarily brief description of the rest of Bohr's life. To whet your curiosity, here is a summary: in 1916, Bohr was appointed professor of theoretical physics at the University of Copenhagen, a position created specifically for him—an enormous recognition for one only about 30 years old. The next year, he began petitioning the University to establish an Institute for Theoretical Physics, which would come into existence (with himself as Director) in 1921, the year before he was awarded the Nobel Prize for Physics. The Institute, which still exists, would become a focal point for both theoretical and experimental physicists in the pre-World-War-II years, with many established physicists and up-and-coming younger researchers and students from around the world visiting or holding longer-term appointments. Bohr became a beloved mentor to many of these people, and would later help many of them find safety and positions abroad as World War II approached. During the 1920s, he worked in conjunction with a number of collaborators to extend his theory to multi-electron atoms, and then became deeply involved in the development of quantum mechanics, particularly questions regarding its meaning and interpretation. With atomic structure becoming understood, forefront research moved to address the structure of nuclei and the properties of nuclear reactions, and here again Bohr remained active. This work culminated in the discovery of uranium fission in late 1938/early 1939, and Bohr played a prominent role in establishing the initial understanding of that phenomenon, being the first to propose that it was the rare isotope uranium-235 that was fissile and could be used to make nuclear weapons. This realization would lead to the enormous work of the Manhattan Project during World War II, and ultimately the bombings of Hiroshima and Nagasaki. Bohr was involved in technical work at the Los Alamos Laboratory during the war, and, foreseeing a postwar arms race, became politically active in attempting to alert both Franklin Roosevelt and Winston Churchill to the potential dangers of nuclear weapons and the necessity for working toward a more open world. These events, along with the dramatic stories of his still-controversial meeting with Werner Heisenberg in the fall of 1941 and his subsequent escape to Sweden (1943) have little to do with his atomic model, but are fascinating in their own rights. Students especially should know some of this history, because it reminds us that the personalities who appear in textbooks are real human beings, with lives full of successes, failures, tragedies, controversies, families, and, ultimately, legacies.

This book is divided into five chapters. No new scientific theory ever comes out of thin air; there is always a background of empirical data and relationships that beg for a physics explanation in the sense of a predictive mathematical model. In the case of the Bohr atom, this background comprised information on the effective sizes of atoms and decades of data in the areas of thermal physics and spectroscopy: Kirchhoff's laws, Max Planck's quantum hypothesis, and the empirical Balmer formula for the wavelengths of the visible spectral lines of hydrogen. Sections 1.1 and 1.2 describe this background. To set the stage for the introduction of the Bohr model, section 1.3 sets up some useful expressions for the forces and energetics involved in the physics of charged particles in circular orbits around each other. A serious problem for all pre-Bohr models of atomic structure was that they predicted that atoms would collapse on timescales of nanoseconds; this is analyzed in section 1.4. To avoid this dilemma, Bohr would have to insert an *ad hoc* assumption into his analysis: science is not always a linear this-hence-that process.

Bohr published his breakthrough theory as a trilogy of papers in the July, September, and November, 1913 editions of *The London, Edinburgh, and Dublin Philosophical Magazine and Journal of Science*. Established in 1798 and now published under the more compact title *Philosophical Magazine*, the 'Phil Mag' is one of the oldest and most venerable scientific journals published in English. The emphasis in this book is on the first of Bohr's papers, wherein he developed his model of so-called 'stationary' energy states and applied it to the hydrogen atom. Chapters 2 and 3 form the heart of this book, presenting derivations of the Bohr model and several of its consequences, as well as descriptions of how the theory was received and extended by other researchers. As with any new theory, some of the reception was critical, and some was enthusiastic. The story begins in section 2.1 with a description of Bohr's early life and how he came to be working with Ernest Rutherford at just the time that Rutherford was developing the concept of the nuclear atom, a tremendous stroke of lucky timing. In his second and third papers, Bohr attempted to treat the structure of molecules and multi-electron atoms, but fuller understandings of the physics of these more complex systems (and in fact of the hydrogen atom as well) had to await the development of wave mechanics in the 1920s, so the discussion of these papers is kept brief.

Chapter 4 deals with a number of supplemental calculations which can be carried out with the Bohr model but which are rarely explored in standard texts. These include isotopic effects in spectral lines of deuterium, estimating the magnetic field created at the nucleus by the orbiting electron, estimating the ionization potentials of two-electron atoms, and recent research on anti-hydrogen and atomic recoils. This latter topic is admittedly not directly a Bohr-model issue, but is mentioned because it has actually been measured. The point of including these calculations is to illustrate that the theory connects to issues well beyond simply explaining the spectrum of hydrogen. Also, a few exercises are embedded in the text at appropriate points.

There exists an enormous historiography on Bohr's life and work; an online search will yield millions of hits. My purpose in this book is not to try to plough new ground in this well-explored area, but rather to offer students a compact bridge to deeper treatments of Bohr, both scientific and humanistic; a list of excellent sources can be found in the Bibliography.

Finally, the appendix gives a summary of physical constants and combinations thereof: they will be useful when doing the exercises.

A final word to students: I hope you enjoy reading this book as much as I enjoyed preparing it. I first encountered the Bohr atom in high school and then as an early-years undergraduate. Later came quantum mechanics proper, with its mystifying wavefunctions, probability maps (see figure 3.9), specialized mathematical functions, and head-spinning interpretations. But Bohr was the foundation; the castle of quantum mechanics was built on his work. Appreciate both.

Cameron Reed
Halifax, Nova Scotia
August 24, 2020

Some selected quotes by and about Bohr

'Bohr's influence on science is only partially expressed in his published work, great as this was. … What he was counted for even more than what he had done'. (George P Thomson, in French and Kennedy 1985, p. xv)

'Perhaps I have found out a little about the structure of atoms'. (To brother Harald in a letter of June 19, 1912, quoted in French and Kennedy 1985, p. 76)

'But, but, but … if anybody says he can think about quantum theory *without* getting giddy it merely shows that he hasn't understood the first thing about it'. (Attributed to Bohr by Otto Frisch in French and Kennedy 1985, p. 136)

Pauli: 'You will probably think that what I said is crazy'. Bohr: 'Yes, but unfortunately it is not crazy enough'. (Exchange between Wolfgang Pauli and Bohr on particle physics, quoted in French and Kennedy 1985, p. 182)

'I am not saying this in order to criticize, but your argument is sheer nonsense'. (Attributed to Bohr by Rudolf Peierls in French and Kennedy 1985, p. 229)

'It can never be done unless you turn the United States into one huge factory' (Bohr's assessment of the possibility of isolating enough uranium-235 to make a nuclear weapon; quoted in Rhodes 1986, p. 294)

'There is no quantum world. There is only an abstract quantum physical description. It is wrong to think that the task of physics is to find out how nature is. Physics concerns what we can say about nature'. (Attributed to Bohr in French and Kennedy 1985, p. 305)

'They built a shining city upon a hill, they who half a century ago founded the quantum mechanics of particles and fields, and the design of this city is understood best through Niels Bohr'. (Philip Morrison in French and Kennedy 1985, p. 350)

'Some things are so serious that one can only jest about them'. (Attributed to Bohr in Pais 1991, p. 510)

References

French A P and Kennedy P J (ed) 1985 *Niels Bohr: A Centenary Volume* (Cambridge, MA: Harvard University Press)

Pais A 1991 *Niels Bohr's Times, in Physics, Philosophy, and Polity* (Oxford: Oxford University Press)

Rhodes R 1986 *The Making of the Atomic Bomb* (New York: Simon and Schuster)

Acknowledgements

I am eternally grateful to all of my teachers, from elementary school on up through graduate school, who instilled in me a love of learning, a love of physics, and who put up with my often clueless questions. Education is an investment where the value is realized years later.

Author biography

Bruce Cameron Reed

Bruce Cameron Reed is the Charles A Dana Professor of Physics Emeritus at Alma College, Michigan. He holds a PhD in Physics from the University of Waterloo in Canada. In addition to a quantum mechanics text, he has published five books on the Manhattan Project, including the IoP Concise Physics volumes *Atomic Bomb: The Story of the Manhattan Project* and *The Manhattan Project: A Very Brief Introduction to the Physics of Nuclear Weapons*. His *Keplerian Ellipses: The physics of the gravitational two-body problem* was also published in the IoP Concise Physics series. In 2009 he was elected a Fellow of the American Physical Society for his contributions to the history of both the physics and the development of nuclear weapons in the Manhattan Project.

IOP Publishing

The Bohr Atom
A guide
Bruce Cameron Reed

Chapter 1

Preliminaries

This chapter summarizes background information relevant to the historical context of the Bohr model. Specifically, we look at what was known of the sizes of atoms, the introduction of quantum concepts to explain phenomena in thermal physics, regularities in spectroscopic data, and the problem of the predicted collapse of atoms in pre-Bohr atomic models. This chapter also sets up some expressions for forces and energies involved in the physics of charged particles orbiting each other. These are useful in analyzing the collapse problem, and set the stage for the derivation of the Bohr atom in chapter 2.

1.1 On the sizes of atoms

An idea of the effective sizes of atoms can be obtained via the following simple argument. Suppose that some element (what is had in mind is a metallic solid) has density ρ grams per cubic centimeter and atomic mass A grams per mole. The number of cubic centimeters per mole of atoms will then be A/ρ. One mole of any kind of entity is Avogadro's number N_A of them, so the volume per atom will be $A/(\rho\ N_A)$. If the atoms are imagined to be packed together tightly as spheres, each will effectively occupy a cube of side length L and volume L^3, as sketched in figure 1.1.

The effective diameter of each atom will be

$$L = \left(\frac{A}{\rho N_A}\right)^{1/3}.$$

Table 1.1 shows results for a few metallic elements drawn from across the periodic table; atomic masses are rounded to the nearest whole number, and values of L are given in Ångstroms (Å). Recall that 1 Å = 10^{-10} m, one ten-billionth of a meter, a unit named after Swedish spectroscopist Anders Ångstrom (1814–74) and convenient as it corresponds closely to the effective sizes of atoms. Despite a range of a

doi:10.1088/978-0-7503-3611-6ch1

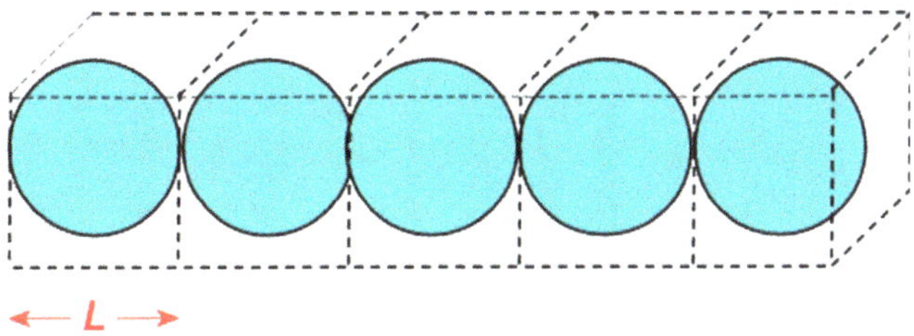

Figure 1.1. Closely-packed atoms occupying cubes of side length L. Imagine extending the structure to three dimensions.

Table 1.1. Effective diameters for various elements.

Element	ρ (g cm^{-3})	A (g mol^{-1})	L (Å)
Li	0.534	7	2.79
Al	2.7	27	2.55
Fe	7.87	56	2.28
Cu	8.96	64	2.28
Pb	11.35	207	3.12
Pt	21.45	195	2.47
U	18.95	238	2.75

factor of over 30 in atomic masses and densities, the effective sizes are remarkably similar: bulk densities and atomic weights indicate that all atoms act as if they are about 2–3 Å in diameter, or ~1–2 Å in radius.

1.2 Spectroscopy, the quantum concept, and the Balmer formula

The history of spectroscopy is extremely rich; only a very brief summary is given here. This section is a bit lengthy, but important for understanding various phenomena and theories which underlay Bohr's atomic model.

That prism-shaped pieces of glass could be used to split beams of sunlight into the constituent colors of the visible spectrum was known in antiquity. Systematic experiments with the solar spectrum began in the 1600s. In particular, Isaac Newton found that upon passing a narrow pinhole-created ray of sunlight through a prism, the spectrum so created could be recombined into white light by passing it through a second prism; flames and stars also served as convenient light sources. Newton believed light to be composed of particulate 'corpuscles', but this idea was superseded in the early 1800s by a wave theory developed by Thomas Young (1773–1829) in order to explain interference effects.

In 1802, English chemist/physicist William Wollaston (1766–1828) built a formal spectrometer that included a slit and lens to focus spectra on a screen for finer examination. This revealed that the colors in the solar spectrum were not distributed continuously, but rather that some colors were missing. This phenomenon came to be called 'dark lines'. This work was picked up by German scientist

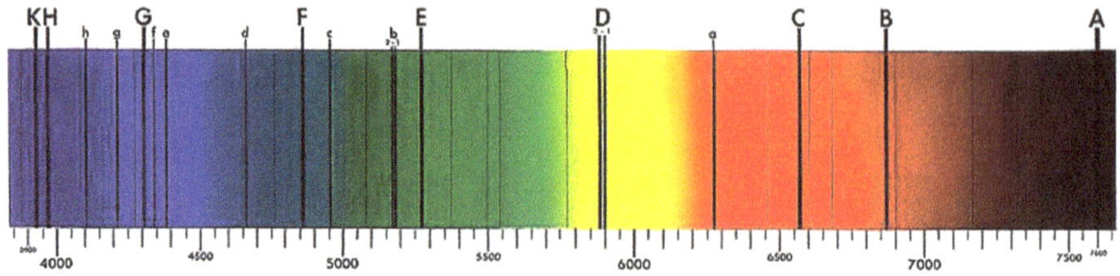

Figure 1.2. Solar spectrum showing prominent Fraunhofer absorption lines. *Source*: Public domain from nl: Gebruiker:MaureenV https://commons.wikimedia.org/wiki/File:Fraunhofer_lines.jpg.

Joseph von Fraunhofer, who is considered to have invented the modern form of the spectroscope around 1814; student laboratory spectrometers still follow his fundamental design. Fraunhofer also refined the development of diffraction gratings, which are essentially a fence of very closely-spaced slits. By knowing the spacing of the slits, it is possible to relate colors to wavelengths via Young's wave theory, and Fraunhofer thus became the founder of quantitative spectroscopy. Fraunhofer cataloged over 500 of Wollaston's dark lines in the solar spectrum; these are now also known as Fraunhofer lines. Today, millions of such lines are known. Figure 1.2 shows some of the Fraunhofer lines in the visible part of the solar spectrum; it was common to refer to particular lines with letters. The scale along the bottom of the figure gives wavelengths in Ångstroms (Å). Modern usage is tending toward using nanometers (nm; 1 nm = 10^{-9} meters = 10 Å), but in this book Ångstroms will be used. Human response to light is a matter of individual physiology, but in general for the visible part of the electromagnetic spectrum, wavelengths run from about 3500–4000 Å for the deepest visible bluish-violet colors to 6500–7000 Å for the reddest.

Chemical spectroscopy flourished with the work of Robert Bunsen (1811–99; of burner fame) and Gustav Kirchhoff (1824–87) in the 1860s. If you have studied electrical circuits, you may have come across 'Kirchhoff's rules' for current conservation at nodes; this is the same Kirchhoff. In the 1860s, this work culminated in what are now known as Kirchhoff's three laws of spectroscopy. These are sketched in figure 1.3, and can be summarized as

(1) An incandescent solid, liquid, or gas under high pressure emits a continuous spectrum: light of all colors appears.
(2) A heated gas under low pressure emits a 'bright line' spectrum: spectral lines at particular wavelengths appear. This is also known as an 'emission' spectrum. The specific lines that appear depend on the gas involved.
(3) When viewed through a cool, low-density gas, a continuous spectrum created by a source as in law (1) produces an 'absorption-line' spectrum: a continuous spectrum with particular wavelengths removed.

The especially striking fact here is that if the same gas is used in (2) and (3), the wavelengths that appear in emission in (2) are identical to those which are absorbed out of the spectrum in (3). Also particularly striking in (2) and (3) is that each element has its own unique pattern of lines. This gave rise to the science of chemical spectroscopy: the ability to identify the presence of elements by their spectra. It was

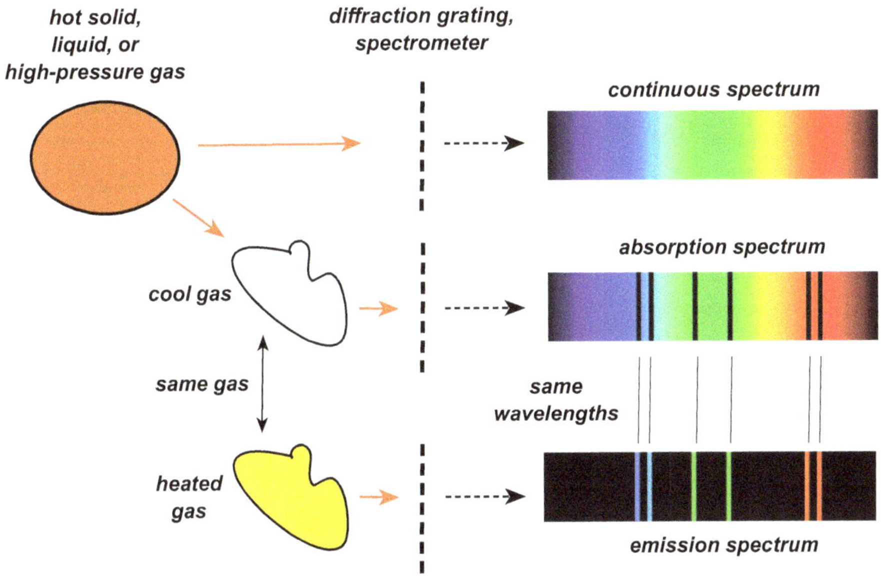

Figure 1.3. Sketch of Kirchhoff's laws of spectroscopy. See also figures 1.4 and 1.5. Colors are approximate. *Source*: Adapted by the author from Talos https://commons.wikimedia.org/wiki/File:Kirchhoff-spectroscopy-law.svg.

not understood why these patterns were present, but they acted like identifying fingerprints of their corresponding elements.

Figure 1.4 shows spectra of a few common elements, with prominent spectral lines indicated by their wavelengths. Figure 1.5 shows the spectrum of hydrogen, which played a key role in Bohr's atomic theory. At visible wavelengths, the hydrogen spectrum comprises a series of lines which get closer and closer together with decreasing wavelength. The reddest line occurs at 6563 Å; the others, strictly an infinite number of them, occur at decreasing wavelengths down to about 3647 Å in the deep violet, where they 'pile up'. (For some qualifications regarding exact wavelengths, see comments associated with table 2.1 in chapter 2.) In practice, spectral lines have finite widths due to phenomena such as the Doppler effect and the effects of the forces that atoms in a gas exert on each other ('pressure broadening'); you cannot separate all of the lines, even with a spectrometer of infinitely great resolution.

An enormous step forward in understanding the nature of light came in 1865 with the publication of James Clerk Maxwell's theory of electromagnetism. For students of physics and electrical engineering, if you have not yet come to Maxwell's equations in your curriculum, you surely will. One of the results of Maxwell's theory was that if an electrical charge is oscillated back-and-forth at some frequency, it will emit energy in the form of an 'electromagnetic wave' which will travel through space at the speed of light, c. This was a quite remarkable development: Maxwell succeeded in marrying all electrical, magnetic, and optical phenomena within one unified theory. All light is a form of electromagnetic radiation.

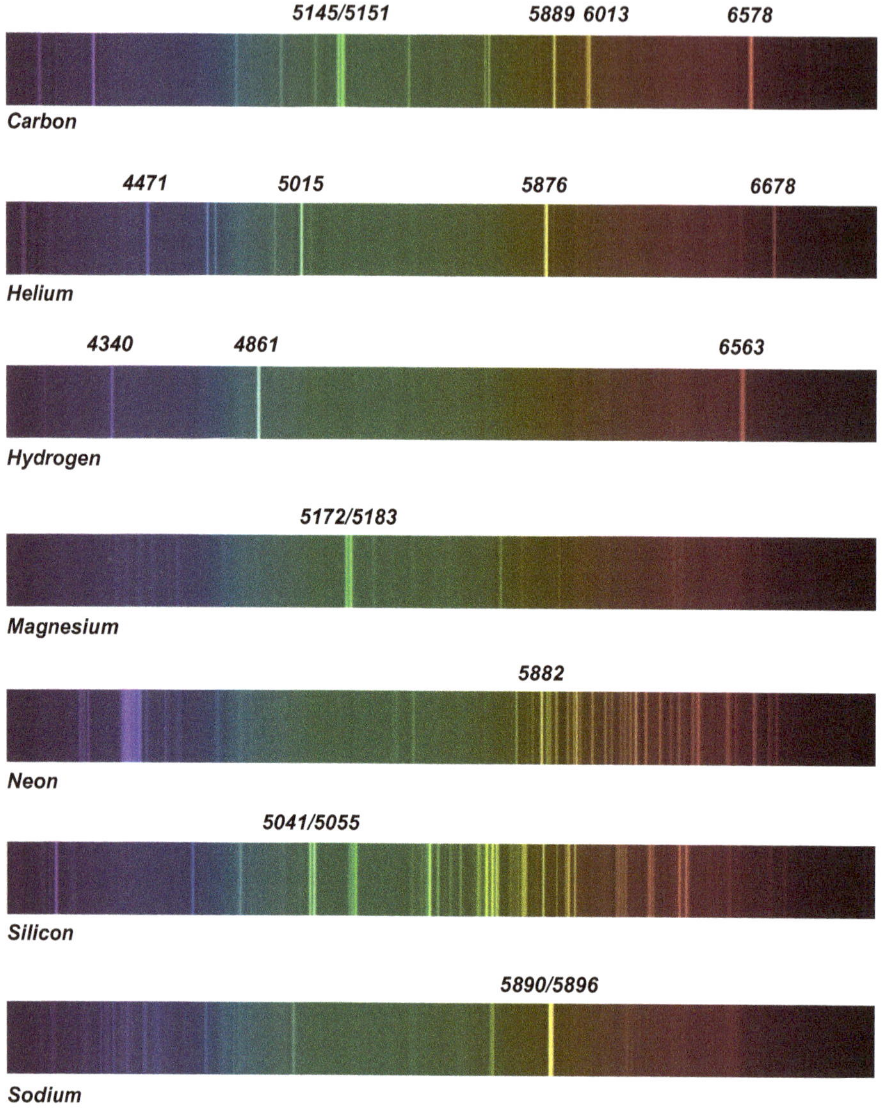

Figure 1.4. Optical spectra of several elements, with wavelengths of some prominent lines indicated in Ångstroms. *Sources*: Public domain from
https://commons.wikimedia.org/wiki/File:Carbon_Spectra.jpg
https://commons.wikimedia.org/wiki/File:Helium_spectra.jpg
https://commons.wikimedia.org/wiki/File:Hydrogen_Spectra.jpg
https://commons.wikimedia.org/wiki/File:Magnesium_Spectra.jpg
https://commons.wikimedia.org/wiki/File:Neon_spectra.jpg
https://commons.wikimedia.org/wiki/File:Silicon_Spectra.jpg
https://commons.wikimedia.org/wiki/File:Sodium_Spectra.jpg.

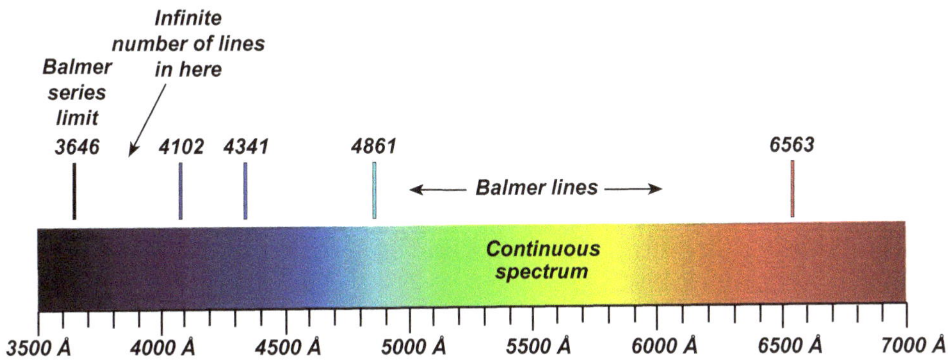

Figure 1.5. Sketch of visible-wavelength continuous spectrum and the first four lines of the Balmer series. Wavelengths are in Ångstroms and are as in air (see section 2.3). Colors are approximate.

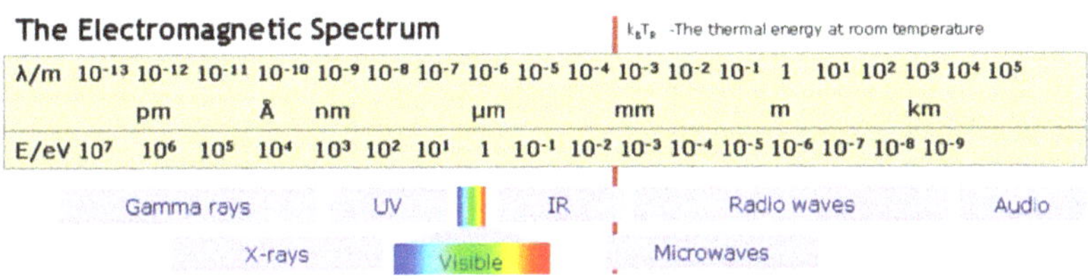

Figure 1.6. The electromagnetic spectrum. The red line at the end of the infrared (IR) part of the spectrum corresponds to room-temperature thermal energy. *Source*: Kristian Molhave https://commons.wikimedia.org/wiki/File:TheElectromagneticSpectrum.jpg. This file is licensed under the Creative Commons Attribution 2.5 Generic license. This illustration was made for the *Opensource Handbook of Nanoscience and Nanotechnology*.

If the frequency of the oscillation is f back-and-forth cycles per second, then the wavelength λ of the emitted wave must be given by the universal wave equation, $\lambda f = c$. In principle, there is no limit to f, so λ is likewise unlimited: it is possible to have an electromagnetic (EM) wave of *any* wavelength. The full run of wavelengths is now known as the *electromagnetic spectrum*. Since atoms within any object that is at any temperature above absolute zero are constantly in motion, they constantly emit EM energy while absorbing it from other emitters. The range of visible wavelengths is only a small segment of the entire EM spectrum: we see only a tiny fraction of the 'light' that is emitted by any object. Figure 1.6 shows a depiction of the EM spectrum and some of the phenomena associated with different wavelengths. Wavelengths are listed along the top of the diagram; the second row gives energies in electronvolts, which we will come to soon. Human beings sense infrared light (IR; beyond red) as heat (think of an open oven door), while shorter-wavelength ultraviolet light (UV) is energetic enough to ionize atoms and molecules in your body and give you a sunburn.

While all objects at temperatures above absolute zero emit energy across the EM spectrum, what figure 1.6 does not depict is that the intensity of the emission is not uniform across all wavelengths. There proves to be a characteristic shape to the intensity/wavelength distribution, which for historical reasons is called a *blackbody* spectrum. This is somewhat far afield from the Bohr atom (although there is an important connection), but there is a point here worth mentioning in order to address a question that might have occurred to you: why do we 'see' in the visible part of the spectrum? It turns out that for objects whose surface temperatures are like that of the Sun (~5800 K), their EM radiation is most intense at greenish-yellow wavelengths. Human beings evolved to take advantage of this; it would not be of much use for us to see at, say, x-ray wavelengths. We might speculate that if intelligent beings have evolved on a planet orbiting a cooler, reddish star, their eyes will involve different chemistry than ours.

Figure 1.7 shows the spectrum of the Sun (jagged white curve) as a function of wavelength in nanometers; the abbreviation AM0 means 'air mass zero', which corresponds to the spectrum as measured outside Earth's atmosphere. The yellow curve corresponds to a 'best-fit' theoretical blackbody spectrum for a temperature of 5778 K. The scale on the y-axis gives the number of Watts of solar energy per square meter per nanometer wavelength interval at the location of the Earth. The intensity peaks at about 500 nm = 5000 Å. Hotter objects have the peak intensity of their blackbody spectra shifted to bluer wavelengths, and cooler ones toward the redder; think of a hot burner on your stove.

A side story on infrared radiation. In 1800, German–British astronomer William Herschel (1738–1822) was testing filters to enable him to observe sunspots, and noticed that a lot of heat seemed to be produced when he was using a filter which let red light pass through. He then directed sunlight to pass through a prism, and held a thermometer up to the projected spectrum, discovering that it read a higher temperature when held off the red end of the spectrum than when placed inside the visible part. This led him to the conclusion that there exist forms of light outside the visible spectrum; he is now considered to have discovered infrared radiation. Sometimes a simple experiment can reveal remarkable information.

Here is a key point that will come up again. Maxwell's theory made it clear that if a charged particle is oscillating at some frequency, it will emit EM radiation of that same frequency, which means a particular wavelength. *If light of particular wavelengths appear in an emission spectrum or do not appear in an absorption spectrum, it*

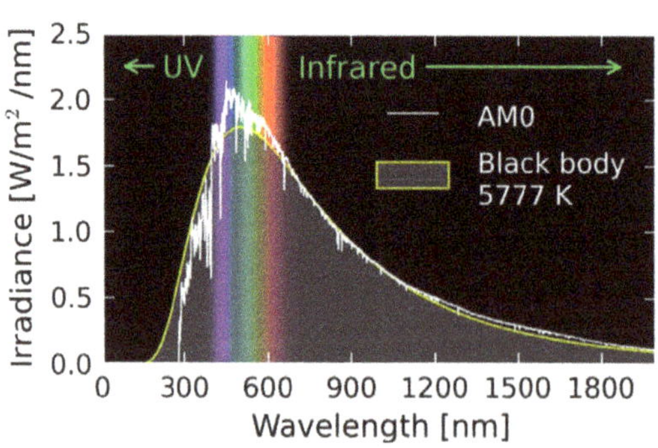

Figure 1.7. Solar spectrum intensity versus wavelength. *Source*: Public domain, Danmichaelo https://commons.wikimedia.org/wiki/File:Solar_AM0_spectrum_with_visible_spectrum_background_(en).png.

means that particular energies are somehow either being emitted or absorbed. This was a crucial clue for Bohr.

Since spectral lines always appear at particular wavelengths, it was natural to assume that something within the structure of atoms which was electrically charged must jiggle back-and-forth at just the right frequencies to emit light detectable in the outside world. With Joseph John 'JJ' Thomson's 1897 discovery of electrons as a universal constituent of matter and the fact that they proved to weigh about 1/2000 as much as individual hydrogen atoms, it was equally natural to presume that atoms comprised clouds of positive electrical material within which were embedded negatively-charged electrons which for some reason were doing the requisite jiggling. The positive cloud was needed to give atoms overall electrical neutrality, and to give them their correct 1–2 Å radii. This Tomson atomic model has often been likened to a plum pudding, with electrons akin to embedded raisins.

The next major step toward the Bohr model occurred in 1900. German physicist Max Planck (1858–1947), whose research area was thermodynamics and statistical mechanics, was attempting to develop a general mathematical expression for blackbody radiation intensity as a function of temperature. At the time, there was no detailed understanding of the internal structure of atoms, so Planck and others idealized them as abstracted 'oscillators' which could vibrate at any chosen frequency and hence emit EM radiation. However, predicted intensity/wavelength relationships based on this scheme never agreed properly with measured values, and the situation was becoming a crisis for an otherwise very successful and well-established branch of physics. Planck found a way around the impasse, but in a way that was regarded as very radical. In classical physics, there is no relationship between the frequency of an oscillator and the amount of energy it possesses. As a rough example, think of the common laboratory experiment where you tie one end of a string to a vibrator which is usually arranged to shake the string up-and-down in a vertical plane, while the other end of the string is fixed, usually by running it over a pulley and tying it to a heavy mass. You can adjust the energy of the vibrator, which dictates its up-and-down amplitude, quite independently of adjusting the frequency of the vibration. This is a mechanical analogy, but the same independence was presumed to hold for atomic-level oscillators and the EM waves which they created. Planck's radical departure was that he found that if he assumed that the energy of an oscillator was in fact restricted to an integer multiple of its frequency, he could reproduce blackbody intensity curves with great accuracy. To make the units work out, frequency had multiplied by a suitable constant, which he designated as h, and which we now know as Planck's constant. Remarkably, *one* universal value of h worked for all temperatures. Planck's constant is now regarded as a fundamental constant of physics.

Expressed mathematically, Planck's assumption was that the energy E of an oscillator of frequency f is given by

$$E = nhf \quad (n = 1, 2, 3, \ldots). \tag{1.1}$$

Frequency has units of 'per second' or s^{-1}, and energy has units of Joules (J), so h must have units of $J/(s^{-1}) = J\ s$, read as 'Joule-seconds' and usually abbreviated to J s. The numerical value, purely empirically, is $h = 6.626 \times 10^{-34}$ J s, a very tiny number. The minuteness of this value means that, on a macroscopic scale where you are dealing with Joules of energy, the possible oscillatory 'energy states' of a system are indistinguishable from each other.

Because E is restricted to multiples of hf through the integer multiplier n, E is said to be 'quantized', that is, it is restricted to particular values for a given f. Frequency itself is still regarded as a continuously-valued variable. Planck is considered to be the father of 'quantum physics' or 'quantum mechanics', a designation he was personally never too comfortable with given his very classical training. Planck was awarded the 1918 Nobel Prize for Physics for his discovery of 'energy quanta'. An excellent (and technically involved) history of the development of the quantum concept can be found in Kuhn (1987).

Planck considered the energy radiated by his oscillators to be classical in that it was still in the form of an EM wave; the idea of electromagnetic radiation as comprising a stream of photons or 'quanta' would not begin to be taken seriously until Einstein's analysis of the photoelectric effect in 1905. None of Planck, Einstein, or Bohr used the term 'photon' in their groundbreaking publications; that terminology would not be coined until 1926, when it was introduced by American chemist Gilbert Lewis of Lewis dot diagram fame (Lewis 1926). However, we can play fast-and-loose with history here to get to a valuable expression that relates energy to wavelength, as in figure 1.6. From the universal wave equation, we can write $f = c/\lambda$. Taking a single photon to correspond to $n = 1$, we can write Planck's formula as

$$E = hf = \frac{hc}{\lambda}. \tag{1.2}$$

When dealing with atomic-scale phenomena, it is customary to quote energies in electronvolts (eV); recall that one eV is the magnitude of the electron charge expressed as a number of Joules of energy: 1 eV = 1.602 … $\times 10^{-19}$ J. Chemical reactions, which involve exchanges of electrons, are typically of energies of a few eV. Equation (1.2) can be put into a handy form for estimating the energy of a photon in eV if its wavelength is specified in Ångstroms. Designate the latter by $\lambda_{Å}$. The wavelength in meters will be $(10^{-10}\ \lambda_{Å})$. In SI units and rounding h and c to three decimal places, we get

$$E = \frac{hc}{(10^{-10}\ \lambda_{Å}\ \text{m})} = \frac{(6.626 \times 10^{-34}\ \text{J s})(2.998 \times 10^{8}\ \text{m s}^{-1})}{(10^{-10}\ \lambda_{Å}\ \text{m})}$$
$$= \frac{1.986 \times 10^{-15}}{\lambda_{Å}}\text{Joules}.$$

Now convert Joules to eV: 1 eV = 1.602×10^{-19} J, or 1 J = 6.242×10^{18} eV. Hence

$$E = \frac{(1.986 \times 10^{-15})(6.242 \times 10^{18})}{\lambda_{Å}}\text{eV} = \frac{12\,400}{\lambda_{Å}}\text{eV}. \tag{1.3}$$

The actual value of the numerator is 12 398.4 Å, but it has been rounded off very slightly for convenience. This is known as the '12 400' formula. For a photon of $\lambda = 5000$ Å, its energy must be ~12 400/5000 ~ 2.48 eV. If we take the range of visible photons to be from 3500 to 7000 Å, the corresponding energy range is from 3.54 down to 1.77 eV. These are much more convenient numbers to deal with than their Joule-enumerated equivalents. Common multiples of the electronvolt are the kilo-electronvolt (1 keV = 1000 eV) and the mega-electronvolt (1 MeV = 10^6 eV); the latter appears frequently in nuclear physics.

On the topic of unit conversions, another one you may run into is that spectroscopists sometimes quote wavelengths *and frequencies* in reciprocal centimeters (cm^{-1}), which means the number of wavelengths that fit within one centimeter. This quantity is also known as the *wave number*. We can arrange the Planck formula to relate energies in eVs and wave numbers as follows. Reciprocal wavelengths are given by

$$\frac{1}{\lambda} = \frac{E}{hc}.$$

If energy is in Joules, then $1/\lambda$ will emerge in reciprocal meters. For energies in eV,

$$\frac{1}{\lambda} = \frac{E_{\text{eV}}(1.602 \times 10^{-19})}{(6.626 \times 10^{-34})(2.998 \times 10^{8})} = (8.065 \times 10^{5})E_{\text{eV}}\ (\text{m}^{-1}).$$

Since 1 m = 100 cm, then one reciprocal meter = 10^{-2} reciprocal cm, so we can convert this to give a result in cm^{-1} by multiplying by 10^{-2}:

$$\frac{1}{\lambda} = 8065E_{\text{eV}}\ (\text{cm}^{-1}). \tag{1.4}$$

An energy of 1 eV corresponds to a wave number of 8065 cm^{-1}. Conversely, a wave number of 1 cm^{-1} corresponds to $E_{\text{eV}} = (1/8065)$ eV = 1.24×10^{-4} eV. Wave number is often designated by the symbol $\tilde{\nu}$, the Greek letter 'nu' (as in *nu*mber) with a tilde (~) over it.

If you know a wavelength directly in Ångstroms, then the wave number in cm^{-1} is just $10^8/\lambda_{\text{Å}}$ because 1 Å = 10^{-8} cm. For example, if $\lambda_{\text{Å}} = 5000$ Å, the wave number is 20 000 cm^{-1}; you should get the same result by taking the corresponding energy, 2.48 eV, and using equation (1.4).

We now leave Max Planck, but a few further comments on the difficulties of his later life are appropriate. In 1944, his house in Berlin was destroyed in a bombing raid; he was living in the country at the time, but lost decades of papers and correspondence. His son Erwin, to whom he was very close, was part of a plot to assassinate Adolf Hitler in July, 1944; the plot failed, and Erwin was captured by the Gestapo and executed by hanging the following January. Planck is now memorialized by several 'Max Planck Institutes' which carry out research in fields across the natural and social sciences.

Stepping back in history, we come to the last piece in the foundations of the Bohr model, a formula published in 1885, the year of Bohr's birth, by

Johann Jakob Balmer (1825–1898; figure 1.8), an otherwise obscure Swiss mathematician (Balmer 1885). Balmer earned a PhD at the University of Basel in 1849, and made his living as a teacher at a secondary school for girls as well as lecturing at his alma mater. His main field of interest was geometry, but he is remembered for a contribution to physics.

Balmer found that, purely empirically, the wavelengths of the lines in the visible part of the hydrogen spectrum could be described by the simple formula

$$\lambda_j = (3646\ \text{Å})\left(\frac{j^2}{j^2 - 4}\right) \quad (j = 3, 4, 5, \ldots). \tag{1.5}$$

Balmer had no physical argument behind this result, it was purely a matter of the formula reproducing measured wavelengths. At the time of his work, over a dozen hydrogen lines had been detected in the spectra of some stars, and his formula fit all of them to within the experimental errors. In the rarefied outer atmospheres of stars, pressure-broadening effects are not as dramatic as in Earth-bound laboratories, and more lines can be photographed and clearly distinguished from each other.

It will be convenient to express Balmer's formula as a prescription for inverse wavelengths, that is, wave numbers. Inverting the formula gives

$$\frac{1}{\lambda_j} = \frac{1}{(3646\ \text{Å})}\left(\frac{j^2 - 4}{j^2}\right) = \frac{4}{(3646\ \text{Å})}\left(\frac{j^2/4 - 1}{j^2}\right),$$

or

$$\frac{1}{\lambda_j} = \frac{4}{(3646\ \text{Å})}\left(\frac{1}{4} - \frac{1}{j^2}\right) = (1.097 \times 10^7\ \text{m}^{-1})\left(\frac{1}{4} - \frac{1}{j^2}\right). \tag{1.6}$$

Figure 1.8. Left: Johann Balmer. Source: https://upload.wikimedia.org/wikipedia/commons/5/51/Balmer.jpeg. Right: Johannes Rydberg. Source https://upload.wikimedia.org/wikipedia/commons/7/7e/JanneRydberg.jpg.

The factor of 1.097×10^7 m^{-1} is now known as the Rydberg constant for hydrogen, abbreviated R_H; a more precise value can be found in equation (2.11). This terminology refers to Swedish spectroscopist Johannes Rydberg (1854–1919; figure 1.8), who in 1888 found that a generalized version of Balmer's formula of the form

$$\frac{1}{\lambda_{n,j}} = R\left(\frac{1}{n^2} - \frac{1}{j^2}\right) \quad (j > n) \tag{1.7}$$

described many, although not all, of the spectral lines of many other elements, with each element characterized by its own value of R (Rydberg 1890). The Balmer–Rydberg formula hinted that there was some regularity to atomic spectra, and hence presumably to atomic structure itself. But it was by no means clear what that structure might be.

1.3 Circular orbit forces and energetics

Bohr's atomic model can be described as a sort of forced marriage between the classical mechanics of circular orbits with electrical forces and some arbitrary adaptations of Planck's quantum hypothesis. In this section, the classical mechanics part of the derivation is set up.

Bohr's model is commonly described as an electron of charge $-e$ in orbit about a nucleus containing Z protons and hence having charge $+Ze$. Z is the common symbol for the atomic number of an element. In his original derivation, Bohr did not worry about the motion of the electron and the nucleus about their common center of mass, that is, he imagined the nucleus to be fixed while the electron orbited it. However, I account for this motion for sake of completeness. Neglecting this effect is equivalent to assuming that the nucleus is infinitely heavy compared to the orbiting electron, and for even the case of hydrogen this is not at all a bad approximation because the lone proton that comprises the nucleus is over 1800 times heavier than the orbiting electron. But there are some cases, such as when dealing with a comparison of transition wavelengths between ordinary hydrogen and so-called 'heavy hydrogen', where the effects of non-infinite-mass nuclei are crucial to the calculations. In many cases I will assume that the nucleus is infinitely heavy compared to the orbiting electron for sake of simplifying a calculation. But it is always easier to start with a general formula and simplify as necessary, rather than to try to go the other way.

Figure 1.9 shows the electron/nucleus system. The electron has mass m_e and executes a circular orbit of radius r_e about the center of mass (CM). The nucleus has mass m_{nuc} and orbital radius r_{nuc} about the CM. The orbital speeds are v_e and v_{nuc}, and are presumed to be such that the electron and nucleus are always directly opposite each other; if they were not, the CM would move.

If we take the CM to define the coordinate origin, then basic mechanics tells us that the masses and orbital radii must be related according as

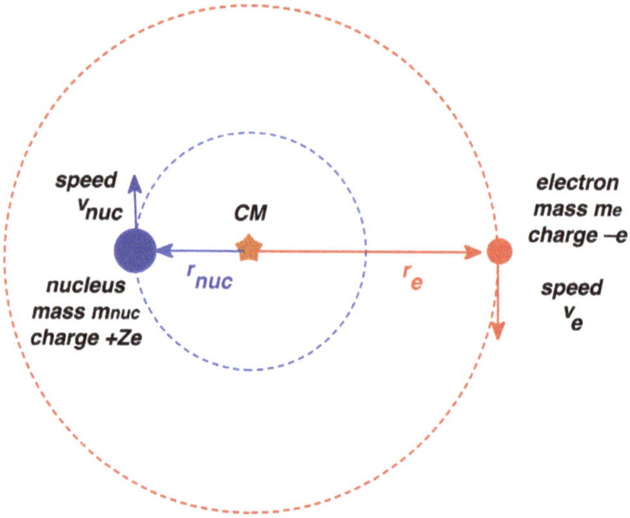

Figure 1.9. Nucleus of mass m_{nuc} and charge $+Ze$ and an electron of mass m_e and charge $-e$ in circular orbits about their mutual center of mass *CM*. Colors are for illustrative purposes only.

$$m_{\text{nuc}} r_{\text{nuc}} = m_e r_e \Rightarrow \frac{r_{\text{nuc}}}{r_e} = \frac{m_e}{m_{\text{nuc}}}. \tag{1.8}$$

The manipulation to form r_{nuc}/r_e will be used in a moment.

For the electron and nucleus to remain opposite each other, they must have the same orbital period. In a circular orbit, the period is given by $2\pi r/v$, so this demands

$$\frac{r_{\text{nuc}}}{v_{\text{nuc}}} = \frac{r_e}{v_e} \Rightarrow v_{\text{nuc}} = \frac{r_{\text{nuc}}}{r_e} v_e. \tag{1.9}$$

Now, the total kinetic energy of the system will be the sum of the individual kinetic energies:

$$KE = \frac{1}{2} m_e v_e^2 + \frac{1}{2} m_{\text{nuc}} v_{\text{nuc}}^2.$$

Replace v_{nuc} with (1.9), and use equation (1.8) for r_{nuc}/r_e. This gives

$$KE = \frac{1}{2} m_e v_e^2 \left[1 + \frac{m_e}{m_{\text{nuc}}} \right]. \tag{1.10}$$

This expression has the advantage of expressing the total kinetic energy of the system in terms of that of the electron alone and the ratio of the two masses. If the nucleus is infinitely heavy, then the *KE* of the system will be that of the electron alone.

It is customary to express this kinetic energy in a modified form by the following manipulation:

$$KE = \frac{1}{2}m_e v_e^2\left[1 + \frac{m_e}{m_{\text{nuc}}}\right] = \frac{1}{2}m_e v_e^2\left[\frac{m_{\text{nuc}} + m_e}{m_{\text{nuc}}}\right] = \frac{1}{2}m_e^2 v_e^2\left[\frac{m_{\text{nuc}} + m_e}{m_e m_{\text{nuc}}}\right].$$

The inverse of the term in square brackets is known as the *reduced mass* of the system. This is always given the symbol μ:

$$\mu = \left[\frac{m_e m_{\text{nuc}}}{m_{\text{nuc}} + m_e}\right]. \tag{1.11}$$

If $m_{\text{nuc}} \gg m_e$, then the denominator of μ reduces to approximately just m_{nuc}, μ reduces to $\sim m_e$, and the kinetic energy again reduces to that of the electron alone. In general, however,

$$KE = \frac{1}{2}\frac{m_e^2 v_e^2}{\mu}. \tag{1.12}$$

In atomic systems, the reduced mass is always very slightly less than the mass of the electron (hence the 'reduced' terminology). For the specific case of hydrogen where the nucleus is a proton (mass m_p), the reduced mass is designated with the symbol μ_H:

$$\mu_H = \left[\frac{m_e m_p}{m_p + m_e}\right] = 9.104425 \times 10^{-31}\ \text{kg} \sim 0.999456\, m_e. \tag{1.13}$$

An exactly analogous quantity arises in celestial mechanics, where the Sun plays the role of the nucleus and a planet that of the electron. Reduced mass was a well-known quantity in classical mechanics long before Bohr's time.

Now, if the electron is moving in a circular orbit, we further know from Newtonian mechanics that it must be subject to a centripetal force of magnitude $m_e v_e^2/r_e$. What provides this force? The only force acting in the system is that of electrical attraction between the electron and the nucleus, which is given by Coulomb's law. Bohr sensibly assumed that this was the cause of the mutual motions, so we put

$$\frac{m_e v_e^2}{r_e} = \frac{Ze^2}{4\pi\varepsilon_o(r_e + r_{\text{nuc}})^2}. \tag{1.14}$$

In writing this expression, we are dealing with *magnitudes*, so the charge of the electron goes into the algebra as $|-e|$. In the denominator on the right side, the inverse-square force acts over the total distance between the electron and the nucleus, $r_e + r_{\text{nuc}}$, but on the left side only the radius of the electron's orbit appears in the centripetal force. Also, it is customary in electrodynamics texts to write the combination $1/4\pi\varepsilon_0$ as just k, but I will leave everything in terms of fundamental physical constants. Also, the resulting formulae look very impressive with all their Greek letters.

Solve (1.14) for the square of the electron's speed:

$$v_e^2 = \frac{Ze^2 r_e}{4\pi\varepsilon_o m_e (r_e + r_{\text{nuc}})^2} = \frac{Ze^2}{4\pi\varepsilon_o m_e r_e (1 + r_{\text{nuc}}/r_e)^2}.$$

In the last step, a factor of r_e was extracted from within the bracket in the denominator to give r_e^2 in the denominator, and then one factor of r_e in this r_e^2 is canceled by the r_e in the numerator. The remaining factor of r_{nuc}/r_e within the bracket can be put in terms of masses via equation (1.8), and the result can be expressed in reduced-mass form as

$$v_e^2 = \frac{Ze^2 \mu^2}{4\pi\varepsilon_o m_e^3 r_e}. \tag{1.15}$$

Substitute this back into equation (1.12) to express the kinetic energy of the system as

$$KE = \frac{1}{2}\frac{m_e^2 v_e^2}{\mu} = \frac{1}{2}\frac{m_e^2}{\mu}\frac{Ze^2\mu^2}{4\pi\varepsilon_o m_e^3 r_e} = \frac{Ze^2\mu}{8\pi\varepsilon_o m_e r_e}. \tag{1.16}$$

Now, electromagnetic theory also tells us that a system of two electrical charges Q_1 and Q_2 separated by a distance d possesses potential energy given by $Q_1 Q_2/4\pi\varepsilon_0 d$. In applying this to our nucleus/electron system, the charges are $+Ze$ and $-e$. Here the negative sign of the electron's charge is built into the algebra explicitly: this is an *energy* calculation, and energy can be positive or negative. Equation (1.14) is a *magnitude-of-force* argument, so the sign of the electron charge disappeared when taking absolute values. For the potential energy,

$$PE = -\frac{Ze^2}{4\pi\varepsilon_o (r_e + r_{\text{nuc}})} = -\frac{Ze^2\mu}{4\pi\varepsilon_o m_e r_e}, \tag{1.17}$$

where the orbital radii were again replaced with masses by extracting a factor of r_e in the denominator and putting things in terms of the reduced mass.

The total energy of the system is the sum of the kinetic and potential energies, which from (1.16) and (1.17) emerges as

$$E = KE + PE = -\frac{Ze^2\mu}{8\pi\varepsilon_o m_e r_e}. \tag{1.18}$$

The negative sign indicates that the system is in a 'bound' energy state. Electrons and protons attract each other, and if they are in some stable configuration, then an input of a positive amount of energy will be required to pull them apart. This energy expression will play an *enormous* role in much of the remainder of this chapter.

So far, there has been absolutely nothing 'quantal' about what has been derived; that will come in chapter 2. For the following section, we return to an issue raised in the Preface.

1.4 The radiative collapse problem

It was remarked in the Preface that one of the great dilemmas of early atomic theories was that they predicted that atoms should collapse almost instantaneously. This section develops an analysis of this issue.

One of the results of James Clerk Maxwell's electromagnetic theory was that an accelerated electrical charge would radiate away energy in the form of electromagnetic radiation. The rate of energy loss depends on the charge and the magnitude of the acceleration a. For an electron, the theory predicted a rate of energy loss given by

$$\frac{dE}{dt} = -\frac{e^2 a^2}{6\pi\varepsilon_0 c^3}, \tag{1.19}$$

where c is the speed of light. This formula was first developed by Irish physicist Joseph Larmor in 1897.

Figure 1.10 sketches the scenario had in mind here: an electron begins with orbital radius R around a nucleus, and then spirals inward due to energy loss, eventually merging with the nucleus in a catastrophic matter/anti-matter explosion. Refer to equation (1.18) for the total energy of such a system: if there is energy loss, E will become progressively more negative, which corresponds to the orbital radius becoming smaller and smaller. The essential question is: if R has a value characteristic of atomic sizes, say ~1 Å, how long will the spiral-in take?

If the spiral-in takes much longer than the characteristic time for one orbit, we can model the electron as being in a circular orbit at any time, with the acceleration being the corresponding centripetal acceleration. The validity of this assumption is investigated *a posteriori*.

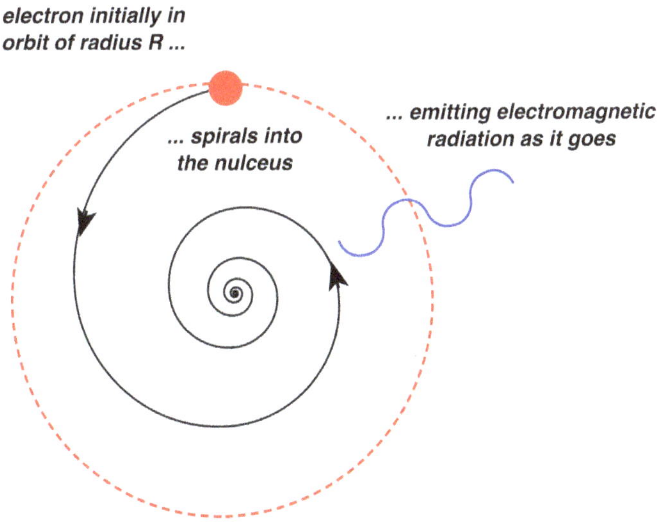

Figure 1.10. An electron spirals into its nucleus, emitting electromagnetic radiation and losing energy as it goes. Colors are for illustrative purposes only.

This is a case where we treat the nucleus as immobile, and write the reduced mass μ as m_e. With equation (1.15), we can then write the acceleration when the electron is at radius r as

$$a = \frac{v_e^2}{r_e} = \frac{Ze^2}{4\pi\varepsilon_o m_e r_e^2}.$$

The rate of energy loss is then

$$\frac{dE}{dt} = -\frac{e^2}{6\pi\varepsilon_o c^3}\left(\frac{Ze^2}{4\pi\varepsilon_o m_e r^2}\right)^2 = -\frac{Z^2 e^6}{96\pi^3\varepsilon_o^3 m_e^2 c^3}\frac{1}{r^4}. \tag{1.20}$$

Now, at any moment, the total energy of the system will be, from equation (1.18),

$$E = -\frac{Ze^2}{8\pi\varepsilon_o r}.$$

Taking the derivative of this with respect to time gives

$$\frac{dE}{dt} = \frac{Ze^2}{8\pi\varepsilon_o r^2}\left(\frac{dr}{dt}\right). \tag{1.21}$$

Putting together equations (1.20) and (1.21) gives

$$\frac{Ze^2}{8\pi\varepsilon_o r^2}\left(\frac{dr}{dt}\right) = -\frac{Z^2 e^6}{96\pi^3\varepsilon_o^3 m_e^2 c^3}\frac{1}{r^4},$$

or

$$\left(\frac{dr}{dt}\right) = -\frac{Ze^4}{12\pi^2\varepsilon_o^2 m_e^2 c^3}\frac{1}{r^2}.$$

Separating variables and integrating from $r = R$ at $t = 0$ to $r = 0$ at $t = t_{\text{collapse}}$ gives

$$\int_R^0 r^2 dr = -\left(\frac{Ze^4}{12\pi^2\varepsilon_o^2 m_e^2 c^3}\right)\int_0^{t_{\text{collapse}}} dt.$$

This gives

$$t_{\text{collapse}} = \left(\frac{4\pi^2\varepsilon_o^2 m_e^2 c^3}{Ze^4}\right)R^3 \sim \frac{(1.05 \times 10^{20}\ \text{m}^{-3})}{Z}R^3.$$

For $R = 1$ Å and $Z = 1$ (hydrogen), the collapse time is $\sim 10^{-10}$ s: *about one-tenth of a nanosecond.* It will probably take you much more than one-tenth of a nanosecond to read this sentence, during which time you have hopefully not collapsed. Clearly, there must be something terribly wrong with this model. The prospect of this incredibly brief collapse time is what prompted Bohr to arbitrarily assume that electrons had to be in some sort of non-radiating states wherein they did not emit electromagnetic radiation even as they orbited the nucleus.

Is the assumption that the spiral-in will occur over many orbits valid? We can check this by comparing the typical orbital period to the spiral-in period. For an orbit of radius R and speed v, the orbital period will be $2\pi R/v$. Using equation (1.15) for the speed (with $\mu = m_e$) gives

$$t_{\text{orbit}} = \frac{2\pi R}{v} = 2\pi R\sqrt{\frac{4\pi\varepsilon_o m_e R}{Ze^2}} = \frac{4\pi^{3/2}\varepsilon_o^{1/2}m_e^{1/2}R^{3/2}}{Z^{1/2}e}.$$

Hence

$$\frac{t_{\text{collapse}}}{t_{\text{orbit}}} = \left(\frac{4\pi^2\varepsilon_o^2 m_e^2 c^3}{Ze^4}\right)R^3\left(\frac{Z^{1/2}e}{4\pi^{3/2}\varepsilon_o^{1/2}m_e^{1/2}}\right)\frac{1}{R^{3/2}} = \left(\frac{\pi^{1/2}\varepsilon_o^{3/2}m_e^{3/2}c^3}{Z^{1/2}e^3}\right)R^{3/2}.$$

The term in brackets evaluates to $2.66 \times 10^{20}\ \text{m}^{-3/2}$; for $R = 1$ Å, $t_{\text{collapse}}/t_{\text{orbit}} \sim 266\,000$. The spiral-in time is much longer than the orbital time, so treating the spiral-in as a succession of circular orbits is plausible.

References

Balmer J J 1885 *Ann. Phys. Lpz.* **261** 80–7

Kuhn T S 1987 *Black-Body Theory and the Quantum Discontinuity 1894–1912* (Chicago, IL: University of Chicago Press)

Lewis G 1926 *Nature* **118** 874–5

Rydberg J R 1890 *Philos. Mag. Ser. 5* **29B** 331–7

Chapter 2

The Bohr model I: hydrogenic atoms

This chapter deals with Bohr's early life and then moves into a detailed derivation of the Bohr atom and how it successfully explained spectral phenomena, particularly in the cases of hydrogen and helium. Also explored are the central role of angular momentum in the theory, Bohr's 'Correspondence Principle', and initial reactions to the theory.

2.1 Niels Bohr: early life and paper I of the trilogy

Partly reproduced from Reed 2019 *The History and Science of the Manhattan Project* (Berlin: Springer) with permission of Springer.

Niels Henrik David Bohr was born on October 7, 1885, in Copenhagen, Denmark (figures 2.1 and 2.2). The Bohr household was a cultured, educated one. His father, Christian, was a physiologist and professor at the University of Copenhagen; his mother, Ellen (née Adler) came from a prominent banking and political family. Niels was the Bohr's second child and first son. Their eldest child, Jennifer Bohr (1883–1933) seems to have left no descendants. Eighteen months after Niels's birth, another son, Harald, came along. Harald became a well-regarded mathematician; both sons were active athletes, with Harald playing for the 1908 Danish Olympic team. Niels and Harald remained close until the latter's death in 1951.

Niels entered the University of Copenhagen in 1903. Recognition of his scientific abilities began early in his career; in 1907, he was awarded the gold medal of the Royal Danish Academy of Sciences and Letters for a prize essay on research into surface tension in vibrating liquid jets; the properties of liquid droplets would figure very prominently in a later phase of his career. For his Master's degree in 1909 and then his doctoral degree (awarded May, 1911), he researched the theory of electron transport in metals, concluding that then-current understanding of atomic structure was inadequate to explain the origin of magnetism.

doi:10.1088/978-0-7503-3611-6ch2

Figure 2.1. Left: Bohr about 1910. Right: Margrethe Nørlund and Niels Bohr at the time of their engagement, 1910. *Source*: (Left) public domain; https://commons.wikimedia.org/wiki/File:Niels_Bohr_-_LOC_-_ggbain_-_35303.jpg; (Right) public domain; https://commons.wikimedia.org/wiki/File:Niels_Bohr_and_Margrethe_engaged_1910.jpg

Figure 2.2. Left: Bohr, 1935. Right: Ernest Rutherford and Niels Bohr. Mrs Mary Rutherford is at the extreme left; the photographer's wife, Rosa Oliphant, is seated in front of Ernest Rutherford. Photograph by Mark Oliphant, courtesy of AIP Emilio Segrè Visual Archives, Margrethe Bohr Collection. *Source*: (Left) public domain; https://commons.wikimedia.org/wiki/File:Niels_Bohr_1935.jpg

In September, 1911, Bohr arrived at Cambridge University to carry out postdoctoral work for a year with J J Thomson. However, the two did not hit it off particularly well, perhaps because of Bohr's criticisms of Thomson's electron theories. But later that year Bohr met Ernest Rutherford, then at Manchester University; they became fast friends, and Bohr transferred to Manchester in March, 1912. Over the next 25 years until Rutherford's death in 1937, Bohr and Rutherford would visit each other on numerous occasions, and ultimately exchanged some 170 letters.

The timing of Bohr's move to Manchester could not have been more propitious. In July, 1911, Rutherford had published his revolutionary discovery that atoms are constructed like miniature solar systems, with the vast majority of the mass of the atom residing in a positively-charged central nucleus, around which were distributed electrons. This new 'nuclear' model of the atom soon displaced J J Thomson's plum-pudding atomic model with its embedded raisin-electrons.

Rutherford was driven to his conclusion based on experiments performed by Hans Geiger (of Geiger counter fame), who in 1909 was looking for a project to occupy undergraduate student Ernest Marsden. For several years, Rutherford had been experimenting with passing alpha-particles ejected in the decays of radioactive elements such as radium and uranium through thin metal foils. Alpha-particles are now known to be helium nuclei stripped of their usual complement of two electrons (another Rutherford discovery, 1909). All of his experience indicated that the vast majority of alphas were deflected by only a very few degrees from straight-line paths as they barreled their way through a foil. This behavior was roughly in line with theoretical expectations: Thomson had calculated that the combination of the size of an Ångstrom-scale positively-charged atomic sphere and the kinetic energy of an incoming alpha (itself also presumably about an Ångstrom in size) would be such that an alpha would typically suffer only a small deflection from its initial trajectory. Deflections of a few degrees would be rare, and a deflection of 90 degrees was expected to be so improbable as to never have any practical chance of being observed. In Thomson's plum-pudding atom, collisions between alphas and atoms were not to be imagined as like those between two billiard balls, but rather more like diffuse clouds of positive electricity passing through each other. The alphas would presumably strike a number of electrons during the collision, but the effect of the electrons' attractive force on the alphas would be negligible due to the vast difference in their masses; electrons played no part in Rutherford's work.

Rutherford suggested to Geiger and Marsden that they check to see if they could observe any large-angle deflections of alphas on passing them through a thin foil of gold leaf. Gold was used as the experimental target because it could be pressed into a thin foil only about a thousand atoms thick. Rutherford fully expected a negative result. Geiger and Marsden counted thousands of scatterings, and found, to their great surprise, that about one alpha in every 8000 was bounced backward toward the direction from which it had come. The number of such occurrences was small, but vastly more than what was expected on the basis of Thomson's atomic model. Rutherford was later quoted as saying that the result was 'almost as incredible as if you had fired a 15-inch shell at a piece of tissue paper and it came back and hit you'.

Geiger and Marsden had begun their experiments in 1909, but so unexpected was their result that it took Rutherford over a year to infer that it meant that the positive electrical material within atoms must be confined to much smaller volumes than had been thought to be the case. The alpha-particles, themselves also nuclei, had to be similarly miniaturized; only in this way could the electrical force experienced by an incoming alpha be intense enough to achieve the necessary repulsion to turn it back if it should by chance approach a target nucleus head-on. The vast majority of alpha nuclei sailed through the foil undisturbed, missing gold nuclei by wide margins, but an occasional one would suffer a head-on collision. The compaction of the positive charge required to explain the scattering experiments was dramatic: down to about a hundred-thousandth of an Ångstrom. But atoms still behaved in bulk as if they were about an Ångstrom or two in diameter. Both results were experimentally indisputable and had to be accommodated. Rutherford's solution was that the vast majority of the mass in atoms must be contained within minute, positively charged nuclei, with the much less massive electrons orbiting at distances of about an Ångstrom away. Thus was born the 'Rutherford atom' (later 'Rutherford–Bohr') of high-school and lower-level undergraduate science classes. Atoms are mostly empty space. A sense of just how empty can be had by thinking of the lone proton that forms the nucleus of an ordinary hydrogen atom as scaled up to being two millimeters in diameter, about the size of an uncooked grain of rice. If this enlarged proton is placed at the center of an American football field, the diameter of the closest electron orbit would be at about the goal lines. While it now a common mental image to have electrons acting like little planets orbiting nuclear 'Suns' with electrical attraction standing in for gravity, the concept was revolutionary in 1911.

The first public announcement of this new model of atomic structure seems to have been made on March 7, 1911, when Rutherford addressed the Manchester Literary and Philosophical Society; this date is often cited as the birthdate of the nuclear atom. The formal scientific publication came in July, and directly influenced Niels Bohr's atomic model. Rutherford did not use the term 'nucleus' in his paper; that nomenclature seems to have been introduced by Cambridge astronomer John Nicholson in a paper published in November, 1911; Nicholson will re-appear a little later in this chapter. The term 'proton' was not introduced until 1920, but was coined by Rutherford himself.

Through the spring of 1912, Bohr worked on various aspects of the theory of the new nuclear atom, trying to understand how it might relate to radioactivity, chemical periodicity, and the stopping of alpha-particles in solids, but apparently he did not consider spectra. In July, he returned to Denmark, where on August 1 he married Margrethe Nørlund, whom he had met in 1909 (figure 2.1). Their marriage would last just over 50 years until Niels's death on November 18, 1962. Together, they would have six sons, four of whom survived to adulthood: Christian (1916), Hans (1918–?), Erik (1920–90), Aage (1922–2009), Ernest (1924–2018), and Harald (1928–38). Tragically, Christian died in 1934 in a sailing accident. Aage Bohr would share the 1975 Nobel prize for physics with Ben Mottelson (Britain) and James Rainwater (United States) 'for the discovery of the connection between collective motion and particle motion in atomic nuclei and the development of the theory of the structure of the atomic nucleus based on this connection'. With his father having been awarded the 1922 Prize 'for his services in the investigation of the structure of atoms and of the radiation emanating from them', Niels and Aage Bohr became one of only four father–son teams to be awarded the Prize for Physics. The other three were William Henry Bragg and William Lawrence Bragg (Britain; both 1915); J J Thomson and George P Thomson (Britain; 1906 and 1937), and Manne Siegbahn and Kai Siegbahn (Sweden; 1924 and 1981).

Bohr was not working in a vacuum. Theories of atomic structure had been actively researched following Thomson's discovery of the electron in 1897, with Thomson himself devoting considerable work to examining the stability of rings of electrons embedded within clouds of positive electricity. Overall, the goal of these models was to try to explain the properties of elements. However, these models often suffered from instabilities if the electrons were imagined to be subjected to slight perturbations, offered no clues to understanding spectra, and were plagued by the radiative collapse problem described in section 1.4. In 1910, Arthur Hass, a doctoral student at the University of Vienna, posited a model for the hydrogen atom where an electron moved on the surface of a sphere of positive electricity, adopting an energy/frequency relationship involving Planck's constant. Remarkably, Haas deduced the same expression that Bohr would in 1913 for what is now called the Bohr radius (the orbital radius of the electron in its lowest-energy state), but he did not consider spectra; this was just *before* the emergence of Rutherford's nuclear model. In 1911 and 1912, John Nicholson published papers in which he attributed spectral lines of stars to various modes of vibration of electrons around equilibrium orbits in the field of a central charge, and posited that the angular momentum of an atom could change by only discrete amounts upon emitting or absorbing energy; angular momentum would prove to be of central importance in Bohr's model (Nicholson 1911). Nicholson's speculations as to what elements corresponded to what atomic numbers were muddled, but between his work and Haas's, clues as to the nature of atomic structure were beginning to emerge. Nicholson was not the first to consider non-Thomson atomic models; in 1904, Japanese physicist Hantaro Nagaoka had developed what he called a 'Saturnian' atomic model, inspired by analyses of the stability of the rings of Saturn, although he abandoned the idea a few years later (Nagaoka 1904). In early 1913, Antoine van den Broek, a Dutch lawyer and amateur scientist, extrapolated from Rutherford's work and the behavior of the increase of atomic weights from one element to the next in the periodic table to formulate the idea that the number of positive central charges in the nuclei of atoms (that is, the number of protons), must increase by one with each step along the table.

During the fall and winter of 1912–13, Bohr, aware of these various developments, continued work on electron-ring atomic structure models, and came to be convinced that in a neutral atom, the number of orbiting electrons must be equal to the number of central positive charges. The key breakthrough, however, occurred in early February, 1913, when a colleague, Hans Hansen, asked Bohr for his insight as to the relationship of the Balmer formula to atomic structure. Whether or not Bohr had previously seen Balmer's formula but had not appreciated its relevance to atomic structure has been debated (he is said to have claimed that he had not seen it), but Hansen's inquiry apparently catalyzed his thinking, for he has been quoted as saying that 'As soon as I saw Balmer's formula, the whole thing was immediately clear to me' (French and Kennedy 1985, p. 43, Pais 1991, p. 144). The pieces of the puzzle must have come together quickly, for the first paper of his trilogy, the now-famous one on the theory of the hydrogen atom, was dated April 5.

As remarked in the Preface, Bohr published his breakthrough theory as a trilogy of papers through the last half of 1913 (Bohr 1913a, b, c). Our immediate concern is with the first of these papers.

2.2 The Bohr model: orbital radii and energy levels

This section and the following one give reconstructions of Bohr's derivation of the orbital energy levels and transition wavelengths for the hydrogen atom.

Bohr's derivation involved two steps. The first is to imagine that a lone electron finds a nucleus containing Z protons to 'bind' to, with the two entering into circular orbits about their common center of mass as was analyzed in section 1.3 above. To do so, the electron will have to lose some energy. To see this, recall equation (1.18) for the total energy of such a system if the final radius of the electron's orbit is r_e:

$$E = -\frac{Ze^2\mu}{8\pi\varepsilon_o m_e r_e}, \tag{2.1}$$

where μ is the reduced mass of the system.

If the electron starts out at a great distance from the nucleus (which on an atomic scale could be just 1 m), then $r_e \sim \infty$, and the energy of the system will be essentially zero. If the electron then 'falls into' an orbit of finite r_e, equation (2.1) tells us that the system energy will be negative: it has *lost* energy. Bohr assumed that, as in classical physics, energy must be conserved. To account for the lost energy, he assumed that in performing its binding, the electron sheds energy in the form of (in modern parlance) some number of photons, each of some frequency f_{photon}. The situation is sketched in figure 2.3.

Bohr described the situation as the electron emitting a 'homogeneous radiation' during the binding, by which he meant electromagnetic radiation of a single frequency. Curiously, Bohr was never comfortable with the concept of light as photons that we now take for granted, even though he was willing to adopt Planck's energy/frequency relationship, as is about to be described.

Bohr assumed that the frequency of each of the emitted photons was exactly one-half of the orbital frequency of the electron in its adopted orbit, that is, one-half of the inverse of the orbital period:

$$\text{photon frequency} = \frac{1}{2}(\text{frequency of electron circular orbit}) = \frac{1}{2}\left(\frac{v_e}{2\pi r_e}\right). \tag{2.2}$$

The factor of one-half might have been inspired by taking the average of the electron's initial and final orbital frequencies.

Bohr now connects his emitted radiation to Planck's blackbody-radiation theory, wherein atoms exchange energy via quanta of electromagnetic radiation. As described in section 1.2, the amount of energy involved is taken to be expressed by nhf, where n is an integer ($n = 1, 2, 3, \ldots$), h is Planck's constant, and f is the frequency of the radiation; see equation (1.1). Bohr later remarked that 'It was in the air to try to use Planck's ideas in connection with such things' (Pais 1991, p. 144)

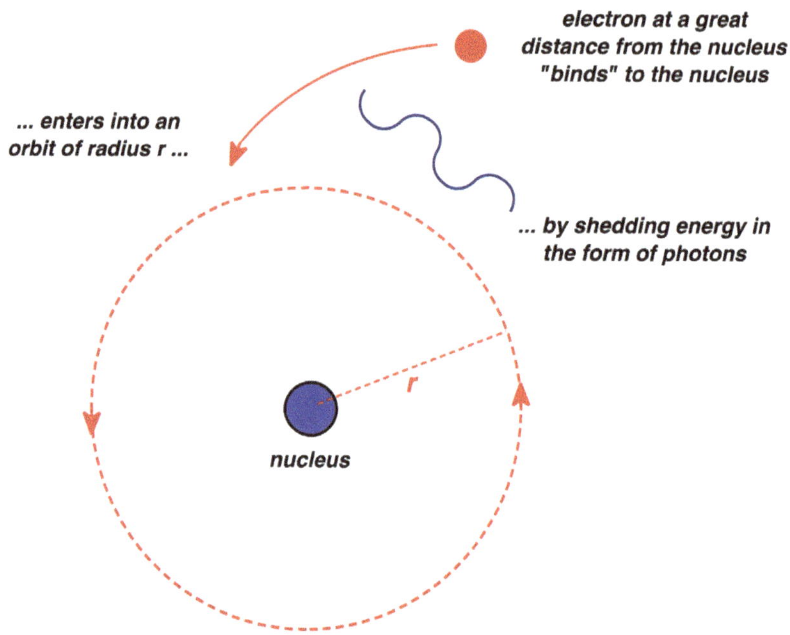

Figure 2.3. An electron 'binds to' a nucleus by losing energy in the form of photons. Strictly, the electron and nucleus enter into circular orbits about their mutual center of mass as in figure 1.9. Colors are for illustrative purposes only.

Bohr took n to specify the number of photons emitted, and then, to conserve energy, set the total energy carried off, nhf_{photon}, to be exactly equal to that which the system had to lose to bring the electron into its final orbit, equation (2.1):

$$nhf_{\text{photon}} = |E| = \frac{Ze^2\mu}{8\pi\varepsilon_o m_e r_e}. \tag{2.3}$$

Using equation (2.2) for the frequency gives

$$nh\left(\frac{v_e}{4\pi r_e}\right) = \frac{Ze^2\mu}{8\pi\varepsilon_o m_e r_e}.$$

The orbital radius cancels in this expression, and what remains can be rearranged to give a formula for the speed of the electron in its orbit,

$$v_e = \frac{Ze^2\mu}{2\varepsilon_o m_e nh}. \tag{2.4}$$

Now recall equation (1.15) for the square of the electron speed that arose in the analysis of the kinetic energy of the electron/nucleus system:

$$v_e^2 = \frac{Ze^2\mu^2}{4\pi\varepsilon_o m_e^3 r_e}. \tag{2.5}$$

Combining this with equation (2.4) gives

$$\frac{Z^2e^4\mu^2}{4\varepsilon_o^2 m_e^2 n^2 h^2} = \frac{Ze^2\mu^2}{4\pi\varepsilon_o m_e^3 r_e},$$

from which it follows that

$$r_e = \left(\frac{\varepsilon_o h^2}{\pi Z m_e e^2}\right) n^2 = \frac{a_o n^2}{Z}. \tag{2.6}$$

This is the first major result of the Bohr theory: *That the electron is restricted to orbits whose radii are squared-integer multiples of* (a_o/Z). The permitted orbits are *quantized.* In classical mechanics, there is no such restriction: you are allowed to imagine an orbit of any radius you please. The integer n is known as the quantum number of the system; what started out as specifying a number of photons has ended up dictating orbital radii. The nucleus would be restricted to its own orbits of much smaller radii as given by equation (1.8), but attention is usually focused on the electron's permitted orbits.

As you might already know, the combination of symbols defining a_o, those in the bracket in equation (2.6) without the factor of Z, is now known as the *Bohr radius. The Bohr radius is dictated entirely by fundamental physical constants.* Even more remarkably, its magnitude corresponds closely to what was determined of the characteristic effective size of atoms in chapter 1:

$$a_o = \left(\frac{\varepsilon_o h^2}{\pi\, m_e e^2}\right) = 5.291772 \times 10^{-11}\,\text{m} = 0.5291772\,\text{Å}$$

At this point, equation (2.6) is put back into the equation for the energy of the system, (2.1), to give

$$E = -\frac{Ze^2\mu}{8\pi\varepsilon_o m_e r_e} = -\frac{Ze^2\mu}{8\pi\varepsilon_o m_e}\left(\frac{\pi Z m_e e^2}{\varepsilon_o h^2}\right)\frac{1}{n^2} = -\left(\frac{Z^2e^4\mu}{8\varepsilon_o^2 h^2}\right)\frac{1}{n^2} \tag{2.7}$$

This is the second major result of Bohr's theory: an explicit expression for the permissible energies ('energy levels') of the system. Remarkably, solving Erwin Schrödinger's later 'wave' equation for the probability distributions of the hydrogen atom gives exactly the same result. The Bohr picture of the atom and quantum-mechanical wavefunctions take very different physical interpretations, but they have this one important result in common. Physicist David Griffiths, author of a highly-respected quantum mechanics textbook, remarks that the Bohr energy formula is '… by any measure the most important result in all of quantum mechanics' (Griffiths 1995, p. 137).

An aspect of equation (2.6) that is easy to overlook while contemplating the success of the Bohr radius is that, unlike the Ångstrom-sized discrete spheres of figure 1.1 which informed the size of Thomson's plum-pudding model, *the Bohr atom has no fixed size*: the permitted orbital radii are unlimited. In principle, the electron could be in an orbit of radius 1 m; indeed, as described in section 2.3.4, radio astronomers routinely detect photons arising from transitions in the region of $n \sim 100$. Such large-scale atoms (on an 'atomic' scale, at least) are now known as Rydberg atoms.

Bohr termed these permitted orbits 'stationary states' in that he arbitrarily assumed that when an electron was in such a state, it for some reason did not radiate any energy, an artifice necessary to avoid the collapse problem of section 1.4. 'Stationary' should be interpreted as meaning 'not radiating', not as 'static'.

Why is this energy formula considered to be so important? In the case of hydrogen ($Z = 1$), the value of the term in brackets is (see equation (1.13) for the reduced mass)

$$\frac{e^4 \mu_H}{8\varepsilon_o^2 h^2} = 2.178\,686 \times 10^{-18}\ \text{Joule} = 13.598\,287\ \text{eV}. \tag{2.8}$$

The great triumph of Bohr's energy formula was that it was known that the greatest amount of energy required to liberate an electron from a proton in a stable hydrogen atom was 13.6 eV, exactly as the formula predicted. This is the 'ionization energy' corresponding to an electron in the $n = 1$ orbital state; higher-n orbits have lower ionization energies. The lowest-energy state, $n = 1$, is known as the 'ground state', and higher-n states are known as 'excited states'.

If you have occasion to study some physical chemistry, you will no doubt encounter a unit of energy called the *Hartree*, named after British physicist Douglas Hartree (1897–1958). The Hartree is used extensively in computational chemistry, and is equal to twice the hydrogenic ground-state energy for $\mu \to m_e$, or about 27.2 eV. Another chemical convention is that energies are often cited in units of calories per mole. The calorie intended here is the 'thermochemical calorie', 1 cal = 4.184 Joules, not the 'food calorie' of a nutrition label, which is 1000 thermochemical calories. The energy of equation (2.8) is 5.207×10^{-19} calories, which for a mole of hydrogen corresponds to about 313 600 calories.

But there was an even more spectacular success of Bohr's analysis about to appear. This corresponds to the second part of argument, and is taken up in the following section.

Exercise

Derive an expression for the centripetal force experienced by the electron in Bohr orbit n as a function of n and physical constants; set $\mu_H = m_e$ and $Z = 1$. What is the value of the force when $n = 1$? Modern force-probe microscopes are capable of measuring nano-Newton scale forces.

[Ans: $F_C = (\pi\, m_e^2\, e^6/4\, \varepsilon_o^3\, h^4\, n^4) = (8.233 \times 10^{-8}/n^4)$ Newtons]

Exercise

What is the orbital frequency of an electron in the $n = 1$ Bohr orbit?

[Ans: $6.58 \times 10^{15}\ s^{-1}$]

Exercise

Both Newton's gravitational law and Coulomb's law are inverse-square laws: the force of attraction between the Sun (S) and the Earth (E) has magnitude GM_SM_E/r^2, while that between the electron and the proton in a hydrogen atom is $e^2/4\pi\varepsilon_o r^2$. This means—in principle—that we can transform the results of Coulomb calculations to gravitational equivalents by substituting GM_SM_E for $e^2/4\pi\varepsilon_o$ wherever the latter appears. Derive an expression for the gravitational equivalent of the Bohr radius; call your result a_G. The masses of the Sun and Earth are $\sim 2 \times 10^{30}$ kg and 6×10^{24} kg, and Earth's orbital radius is about 1.5×10^{11} m. Taking $r = a_G n^2$, what is the 'quantum number' of Earth's orbit? By computing dr/dn, estimate dr for $dn = 1$. Would the change dr be detectable?

[Ans: $a_G = (h^2/4\,\pi^2 G M_S M_E^2)$. $n \sim 2.5 \times 10^{74}$; $dr \sim 10^{-63}$ m: undetectable.]

2.3 The Bohr model: transitions and spectral series

The second part of Bohr's argument leads to a result that is perhaps even more impressive than his energy formula: an explanation of the otherwise purely empirical Balmer formula for the visible-wavelength spectral lines of hydrogen.

Once the electron has completed its binding process, it will be in some orbit n about the center of mass of the system as given by equation (2.6). Suppose that $n > 1$. Now Bohr considers the energetics of the electron *transiting* to another orbit. For the sake of argument, imagine that the electron decides, for some reason, that it wishes to occupy an orbit of lower n value. Label the starting value of n as n_{initial}, and the ending value n_{final}. We assume that $n_{\text{initial}} < n_{\text{final}}$; the opposite case will be discussed presently. Even if there are orbits between these two, it is imagined that the transition takes place in one step, as sketched in figure 2.4. Note that this sketch is by no means to scale: according to equation (2.6), the orbits space according as n^2, so in a scale drawing they would be spacing increasingly farther apart from each other with increasing n.

Look at equation (2.7). If $n_{\text{initial}} > n_{\text{final}}$, then $E(n_{\text{final}})$ is more negative than $E(n_{\text{initial}})$, that is, the system must lose energy to accomplish the transition. Here is where Bohr's second use of Planck's quantum hypothesis enters the picture: he posited that, to accomplish the transition, the electron loses energy in the form of a photon, with the amount of energy corresponding to exactly the difference in energy between the two orbits, $\Delta E_{\text{system}} = (E_{\text{final}} - E_{\text{initial}})$. The wavelength of a photon is

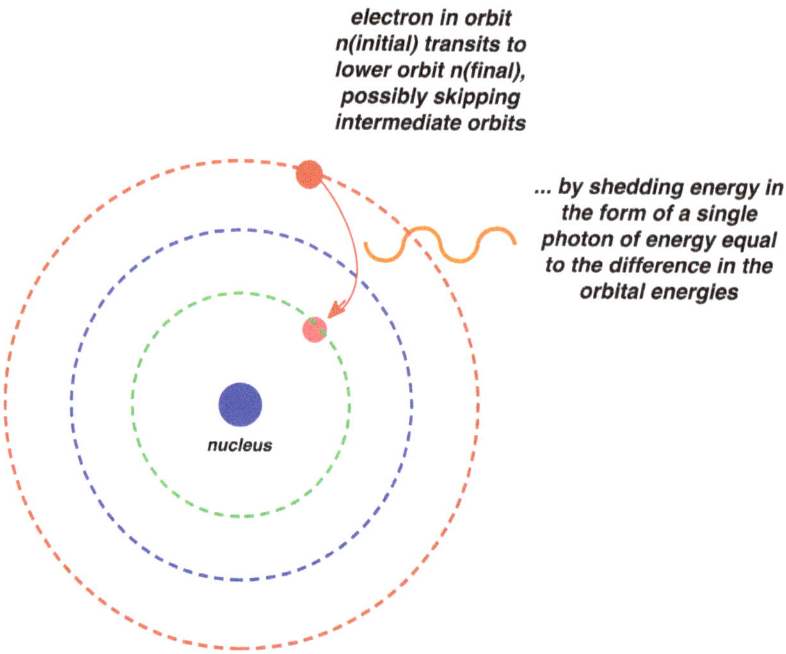

Figure 2.4. An electron undergoes a transition from orbit n_{initial} to orbit n_{final}. In this sketch, $n_{\text{initial}} > n_{\text{final}}$. To accomplish the transition, the system must lose energy. Colors are for illustrative purposes only.

dictated by its energy according as $E = hc/\lambda$ as was used above. As a reminder that this photon is emitted as a consequence of a transition, the subscript $i \to f$ is used:

$$E_{\text{photon}} = \frac{hc}{\lambda_{i\to f}} = -\Delta E_{\text{system}} = -(E_{\text{final}} - E_{\text{initial}}).$$

The negative sign appears in front of ΔE_{system} because energy must be conserved: $\Delta E_{\text{system}} + E_{\text{photon}} = 0$. This subscript $i \to f$ may seem redundant, but it serves as a reminder that these transition-generated photons involve *two* energy levels (orbits), not just one. Hence (be careful with the negative sign):

$$\frac{1}{\lambda_{i\to f}} = \frac{(E_{\text{initial}} - E_{\text{final}})}{hc} = \left(\frac{Z^2 e^4 \mu}{8\varepsilon_o^2 h^3 c}\right)\left(\frac{1}{n_{\text{final}}^2} - \frac{1}{n_{\text{initial}}^2}\right). \quad (2.9)$$

Now you will need to take a deep breath.

Recall the empirical Balmer formula from chapter 1, equation (1.6):

$$\frac{1}{\lambda_j} = R_{\text{H}}\left(\frac{1}{4} - \frac{1}{j^2}\right) = (1.097 \times 10^7 \text{m}^{-1})\left(\frac{1}{4} - \frac{1}{j^2}\right) \quad (j = 3,4,5, \ldots). \quad (2.10)$$

Equations (2.9) and (2.10) are very similar to each other. From the rightmost bracketed term in each, it looks as if the Bohr formula becomes the Balmer formula if $n_{\text{final}} = 2$ and $n_{\text{initial}} = j = 3, 4, 5, \ldots$. But what about the Rydberg constant? If the two formulae are to agree exactly, then Bohr has a prediction: that for hydrogen with reduced mass μ_H, the Rydberg constant must be $R_H = (Z^2 e^4 \mu_H / 8\varepsilon_o^2 h^3 c)$. *There are absolutely no adjustable parameters in this expression: If it does not give the observed value of the Rydberg constant, the Bohr theory must be wrong.* Using the values of the physical constants given in the Appendix, the reduced mass of equation (1.13), and setting $Z = 1$ for hydrogen, we get

$$R_H = \frac{e^4 \mu_H}{8\varepsilon_o^2 h^3 c} = 10\,967\,758 \text{ m}^{-1}, \tag{2.11}$$

in agreement with the observed value! This is a remarkable result: a prediction of an otherwise empirical number in terms of fundamental physical constants alone. For its strange but fruitful combination of classical and quantum concepts, Bohr's model is often described as 'semi-classical'.

To be honest, I have cheated a bit here. In his original publication, Bohr did not worry about the reduced mass, and set $\mu_H = m_e$—but the agreement is still excellent. Bohr's analysis corresponded to assuming an infinite-mass nucleus, in which case the Rydberg constant is adorned with an infinity subscript:

$$R_\infty = \frac{e^4 m_e}{8\varepsilon_o^2 h^3 c} = 10\,973\,731 \text{ m}^{-1}. \tag{2.12}$$

This version of the Rydberg constant is now called *the* Rydberg constant, and it has been measured to an incredible level of precision with modern spectroscopy. Based on measurements involving a cryogenic bream of hydrogen atoms, Beyer *et al* (2017) reported it to an astounding 14 significant figures of accuracy: 10 973 731.568 076 ± 0.000 096 m^{-1}.

For a one-electron atom of an element with atomic number Z and reduced mass μ_Z, the Rydberg constant will be the first bracketed term in equation (2.9):

$$R_Z = \frac{Z^2 e^4 \mu_Z}{8\varepsilon_o^2 h^3 c}. \tag{2.13}$$

Animations of transitions in the Bohr model can be found in various online sources, such as one prepared by the publishers of the *Encyclopedia Britannica* at https://www.britannica.com/video/213886/Overview-Niels-Bohr-refinement-Rutherford-atomic-model.

Several comments can be made on the Balmer–Bohr formula. These are discussed in the following subsections.

2.3.1 The Balmer series

According to the Bohr model, the Balmer series of hydrogen lines corresponds to electrons transiting from orbits with $n > 2$ down to $n_{\text{final}} = 2$.

2.3.2 Emission and absorption

Imagine figure 2.4 in reverse: an electron transits from a lower-n orbit to a higher-n one: $n_{\text{final}} > n_{\text{initial}}$. This would require an input of energy, and this energy would be of exactly the same magnitude as that liberated if the transition were downward between the same two orbits. How could this energy be provided? If there is a flux of photons from the outside world—say from the continuous spectrum emitted by a blackbody—then those photons of exactly the wavelength $1/\lambda_{i\to f}$ will have just the right amount of energy, if they are captured by an electron, to cause the upward transition. These photons will be *removed* from the incident flux: an absorption spectrum. Thus, downward transitions create emission spectra, while absorption spectra are explained as upward transitions. Bohr's model explained not only the ionization energy of hydrogen, the Balmer series, and the Rydberg constant, but also why emission and absorption spectra of a given element always involve the same wavelengths!

2.3.3 Other hydrogen series

Unlike with the Balmer formula, in the Bohr formula the orbit numbers n_{initial} and n_{final} are unrestricted, except for the interpretation of having $n_{\text{final}} < n_{\text{initial}}$ for an emission spectrum and $n_{\text{final}} > n_{\text{initial}}$ for an absorption spectrum. Thus, there is an infinitude of possible (n_{initial}, n_{final}) pairs, each with a corresponding transition wavelength given by equation (2.9) with the reduced mass for hydrogen:

$$\frac{1}{\lambda_{i\to f}} = \left(\frac{e^4\mu_{\text{H}}}{8\varepsilon_o^2 h^3 c}\right)\left(\frac{1}{n_{\text{final}}^2} - \frac{1}{n_{\text{initial}}^2}\right) = (911.76\text{Å})^{-1}\left(\frac{1}{n_{\text{final}}^2} - \frac{1}{n_{\text{initial}}^2}\right), \tag{2.14}$$

and with energies given by

$$E_{\text{photon}} = -(E_{\text{final}} - E_{\text{initial}}) = (13.598\text{ eV})\left(\frac{1}{n_{\text{final}}^2} - \frac{1}{n_{\text{initial}}^2}\right). \tag{2.15}$$

For the sake of explicitness, imagine that we are dealing with emission spectra. Tables 2.1 and 2.2 tabulate the first few pairs of wavelengths (in Ångstroms) and energies (in eV).

Table 2.1. Hydrogen transition wavelengths (Å); equation (2.14).

	n (initial)						
n (final)	2	3	4	5	6	Series limit (∞)	Series
1	1215.7	1025.7	972.5	949.8	937.8	911.8	Lyman
2		6564.7	4862.7	4341.7	4102.9	3647.1	Balmer
3			18 756.3	12 821.7	10 941.2	8205.9	Paschen
4				40 522.8	26 258.8	14 588.2	Brackett
5					74 598.8	22 794.1	Pfund
6						32 823.5	Humphreys

Table 2.2. Hydrogen transition energies (eV) for various (n_{initial}, n_{final}) pairs; equation (2.15).

	n (initial)						
n (final)	2	3	4	5	6	Series limit (∞)	Series
1	10.20	12.09	12.75	13.05	13.22	13.60	Lyman
2		1.89	2.55	2.86	3.02	3.40	Balmer
3			0.66	0.97	1.13	1.51	Paschen
4				0.31	0.47	0.85	Brackett
5					0.17	0.54	Pfund
6						0.38	Humphreys

All possible transitions that terminate at a given value of n_{final} from higher values of n are said to comprise a *series* of lines. For hydrogen, the first six of these are now named after their discoverers, giving Balmer credit for 'discovering' the $n_{\text{final}} = 2$ series. The transition to $n_{\text{final}} = n_{\text{initial}} - 1$ always gives the longest wavelength of a series, and that corresponding to $n_{\text{initial}} = \infty$ is said to correspond to the *series limit*: the shortest possible wavelength for the series.

The Lyman series of lines lies entirely in the ultraviolet part of the spectrum, and is named after American spectroscopist Theodore Lyman IV (1874–1954), who discovered them between 1906 and 1914; that is, some of them were discovered *before* the advent of the Bohr theory, and their wavelengths proved to be exactly as the Bohr model predicted (Lyman 1914). The Paschen series, which lies in the infrared region, is named after the German physicist Friedrich Paschen (1865–1947), who observed them in 1908, again before the development of the Bohr theory; infrared wavelengths are considered to start at ~7000 Å (Paschen 1908). The Brackett series (far infrared) is named after another American, Frederick Brackett (1896–1988), who published his discovery in 1922 (Brackett 1922). The Pfund series is named in honor of August H Pfund (1879–1949), also American, who published in 1924 (Pfund 1924). Finally, the Humphreys series is named after its discoverer, Curtis Humphreys (1898–1986), also American, who detected them in 1953 (Humphreys 1953). Hydrogen series beyond $n_{\text{final}} = 6$ are not given specific names.

A point to note here is that the electron can execute a transition between, say, the $n = 1$ level to the $n = 3$ level without stopping at $n = 2$. A baseball analogy has been made to this: that a runner can get from first to third without touching second (Ellison and Hollingsworth 1976). This raises the deeper question of how does the electron decide when it is going to transit, and to what energy level? Answers to these questions turned out to be another aspect of wave mechanics, and are inherently probabilistic.

A quirk of nature deserves a comment here. Notice that the longest wavelength of the Lyman series falls below the shortest wavelength of the Balmer series; similarly, the longest wavelength of the Balmer series lies below the shortest wavelength of the Paschen series. However, beyond this, this pattern breaks down: the longest wavelength of the Paschen series overlaps into the short-wavelength end of the

Brackett series. What a remarkable stroke of luck for Balmer and Bohr: that the simplest element in nature has a spectral series lying almost exclusively in the visible part of the spectrum, 'unentangled' from its neighbors! It turns out that *only* the Lyman and Balmer series are so separated from their neighbors; this is examined in more detail in section 3.4.

If you check tabulations of standard wavelengths in physics or astronomy texts, you might wonder why they quote the wavelength of the Balmer $n = 3 \rightarrow 2$ transition as typically 6563 Å, or more precisely, 6562.8 Å, as opposed to the 6564.7 Å in table 2.1. The reason for this is that those sources quote wavelengths as measured *in air*, while those in table 2.1 correspond to what would be measured in a vacuum. The index of refraction of air is what causes the difference. In principle, the correction is $\lambda_{air} = \lambda_{vac}/n_{air}$, but in practice things are more complicated: the index of refraction is a function of pressure, humidity, and wavelength itself! For dry air at standard atmospheric pressure and temperature (20°C), $n_{air} \sim 1.000\ 27$ at $\lambda \sim 6330$ Å.

2.3.4 Transition notation

If you have studied more advanced quantum mechanics, you will know that the precise energy levels of electrons involve modifications to the simple Bohr formula for effects such as orbital angular momentum and spin quantum numbers. The result of this is that energy levels become designated with notations such as $4P_{3/2}$ rather than just a simple $n = 4$. For the purposes of this book, however, we will stick with the Bohr model, except for a brief foray into a two-quantum-number elaboration in section 3.8.

A number of notations are used to designate electron transitions. We could, for example, use just $3 \rightarrow 2$ as above to designate $n_{initial} \rightarrow n_{final}$. A more compact notation is to use the symbol for the element involved, followed immediately by the numerical value of n_{final}, followed by a Greek letter: α, β, γ, δ, …. The Greek letters go in the order of increasing possible values of $n_{initial}$ in the order $n_{final} + 1$, $n_{final} + 2$, $n_{final} + 3$, etc. In this scheme, the $5 \rightarrow 3$ transition in hydrogen is designated as H3β. In the specific case of hydrogen, one sometimes sees an abbreviated version of series name, particularly in the case of the Lyman or Balmer series (Ly, Ba), followed by the relevant Greek letter: Lyγ corresponds to the $4 \rightarrow 1$ transition.

In interstellar space, hydrogen atoms with electrons in very high n states can form because the low density of the environment ensures that they will not be immediately ionized by collisions with other atoms. In radio astronomy, the H109α transition ($n = 110 \rightarrow 109$) is an important diagnostic of the conditions in star-forming regions. This transition was first detected by Höglund and Mezger (1965); their receiving equipment was most sensitive at such wavelengths. Radio astronomers prefer to discuss transitions in terms of frequencies for convenience when tuning their equipment. Since $\lambda f = c$, we can use equation (2.14) to write a transition frequency by multiplying by c. For hydrogen,

$$f_{i \rightarrow f} = \frac{c}{\lambda_{i \rightarrow f}} = \left(\frac{e^4 \mu_H}{8 \varepsilon_o^2 h^3}\right)\left(\frac{1}{n_{final}^2} - \frac{1}{n_{initial}^2}\right). \tag{2.16}$$

Putting in the numbers gives

$$f_{i \to f} = (3.288 \times 10^{15}\ \text{s}^{-1})\left(\frac{1}{n_{\text{final}}^2} - \frac{1}{n_{\text{initial}}^2}\right).$$

A reciprocal second is known as a Hertz (Hz), with common multiples of kiloHertz (kHz, 1000 Hz), MegaHertz (MHz, 10^6 Hz), GigaHertz (GHz, 10^9 Hz), and occasionally TeraHertz (THz, 10^{12} Hz). FM radio transmissions, for example, are usually assigned to within a range of 87.5–108 MHz, with frequencies spaced in multiples of 0.1 MHz; the exact spectrum range varies from country to country. If your favorite station advertises itself as broadcasting at 100.5, this means 100.5 MHz.

Exercise

What is the frequency of the H109α transition?

[Ans: ~5009 GHz]

2.4 Ionized helium

In principle, Bohr's model applies to an atom of any element which has been ionized of all but one electron: once-ionized helium, twice-ionized lithium, thrice-ionized beryllium, and so forth. A time-honored application is to look specifically at the case of helium, which Bohr considered in the first paper of his trilogy.

A brief history lesson. Helium was discovered in the spectrum of the Sun during an eclipse in August, 1868, by French astronomer Jules Janssen, who recorded a wavelength of 5875 Å. This is close to that of a pair prominent spectral lines of sodium at around 5890 Å (figure 1.4), and Janssen attributed the new feature to that element. Later the same year, British astronomer Norman Lockyer observed the same feature in the solar spectrum, but concluded that it must be due to a previously-unknown element. Both Janssen and Lockyer are now credited with the discovery. Lockyer and British chemist Edward Frankland dubbed the new element helium, after the Greek word for the Sun, *helios*.

Helium was not detected terrestrially until 1882, when it was found to be present in lava erupted by Mount Vesuvius in Italy. In 1895, Scottish chemist William Ramsey detected it in uranium compounds. This would prove a major clue to its origin. Helium is but a trace gas in Earth's atmosphere; it is chemically inert, and so light that it diffuses off into space. (Atomic hydrogen does so as well, but hydrogen binds with many other elements such as oxygen and carbon, and so remains with us in a multitude of compounds.) The major source of helium is as a by-product of oil and gas wells: it comes from within the Earth, and emerges in gaseous form. But how did it come to be there? Uranium and other heavy elements constantly undergo radioactive decay, emitting alpha particles as they do so. In 1909, Ernest Rutherford and Thomas Royds trapped alpha particles from the decay of radon gas and subjected them to spectroscopic analysis, finding that their spectra were identical to that of helium. Strictly, alpha particles are nuclei of helium-4 atoms, containing two protons and two neutrons. When they are emitted in radioactive decays, the alphas emerge without any attendant electrons, but they quickly pick up free electrons and become neutralized. Every time you fill a balloon with helium or use liquid helium in a superconducting cryogenic instrument, you are using a substance born of radioactive decay.

Chemists use a Roman-numeral system of designating possible ionization states of elements. The neutral form of the element, that is, where the number of electrons is equal to the number of protons, is designated with the symbol for the element followed by the numeral I. If one electron has been removed, the symbol is followed by II; if two electrons have been removed, then III is used, and so forth. In this scheme, carbon atoms which have been stripped of five of their normal complement of six electrons are designated as C VI. Our concern is with He II.

The mass of a helium nucleus is $6.644\,657 \times 10^{-27}$ kg. Designating the mass on an alpha particle as m_α, the reduced mass for our once-ionized helium is, from the general recipe of equation (1.11)

$$\mu_{\text{He}} = \left[\frac{m_e m_\alpha}{m_\alpha + m_e}\right] = 9.108\,135 \times 10^{-31}\,\text{kg} \sim 0.999\,863\, m_e,$$

practically identical to although slightly greater than that for hydrogen, equation (1.13).

With the help of equation (2.13), the Rydberg constant for He II is then

$$R_{\text{He}} = \left(\frac{Z^2 e^4 \mu_{\text{He}}}{8\varepsilon_o^2 h^3 c}\right) = 43\,888\,909\, m^{-1} \tag{2.17}$$

This is essentially four times the Rydberg constant for hydrogen. From equation (2.14), the transition wavelengths are, upon using the appropriate Rydberg constant,

$$\lambda_{i\to f} = R_{\text{He}}^{-1}\left(\frac{1}{n_{\text{final}}^2} - \frac{1}{n_{\text{initial}}^2}\right)^{-1} = (227.85\,\text{Å})\left(\frac{1}{n_{\text{final}}^2} - \frac{1}{n_{\text{initial}}^2}\right)^{-1}. \tag{2.18}$$

He II transition wavelengths in Ångstroms for the first few values of n are listed in table 2.3. Only one series, that for $n_{\text{final}} = 4$, is given a formal name: the Pickering–Fowler series. The 5875 Å of Janssen and Lockyer does not appear in the table because it is a neutral-helium transition.

If you compare tables 2.1 and 2.3, you will see a pattern: that some of the ionized helium transitions are very close in wavelength to some of those in hydrogen. For example, the $6 \to 4$ transition for He II is very similar to the $3 \to 2$ transition for H; similarly, the He II $4 \to 2$ transition is similar to the H $2 \to 1$ transition. This is a quite general pattern: for a hydrogen transition $n_{\text{initial}} \to n_{\text{final}}$, the wavelength of the helium transition $2n_{\text{initial}} \to 2n_{\text{final}}$ is practically identical; Bohr remarked on this in the first of his papers.

Table 2.3. He II transition wavelengths (Å) for various (n_{initial}, n_{final}) pairs; equation (2.18).

	n (initial)						
n (final)	2	3	4	5	6	Series limit	Series
1	303.8	256.3	243.0	237.3	234.4	227.8	
2		1640.5	1215.2	1085.0	1025.3	911.4	
3			4687.2	3204.1	2734.2	2050.6	
4				10 126.6	6562.0	3645.6	Pickering–Fowler
5					18 642.1	5696.2	
6						8202.5	

This factor-of-two pattern can be seen more clearly by extracting the factor of $Z^2 = 4$ from the Rydberg constant in (2.17) and joining it up with the factors of n^2 in (2.18):

$$\lambda_{i\to f} = \left(\frac{8\varepsilon_o^2 h^3 c}{e^4 \mu_{\text{He}}}\right)\left[\left(\frac{2}{n_{\text{final}}}\right)^2 - \left(\frac{2}{n_{\text{initial}}}\right)^2\right]^{-1}.$$

Since the reduced masses in the two cases are very similar, the round-bracketed prefactor is practically identical for the two species. The wavelengths agree for 'doubled' values of the hydrogen transition *n*'s.

Like its neutral counterpart, the spectrum of ionized helium has an astronomical pedigree. Lines of ionized helium were detected in the high-temperature star ζ (zeta) Puppis in 1896 by American astronomer Edward Pickering, who attributed them to a variant form of hydrogen that possessed half-integer transition levels in the Balmer–Rydberg formula. William Fowler, a British astronomer, managed to produce similar lines in a mixture of hydrogen and helium, and supported Pickering's conclusion. Upon publication of his atomic model, however, Bohr argued that they were due to ionized helium. By 1915, spectroscopists and astronomers had come to agree with Bohr's conclusion. Transitions down to $n = 4$ are now recognized as comprising the Pickering–Fowler series.

Wavelengths can be measured with remarkable precision via spectroscopy, and one of the first objections to Bohr's theory was that its predictions of hydrogen and helium wavelengths were not quite accurate. Bohr dispatched this objection in late 1913 in a paper published in *Nature* wherein he modified the theory to take into account the reduced mass of the electron/nucleus system, a feature he had initially neglected (Bohr 1913d). This completely removed the discrepancy, impressing the spectroscopic community.

2.5 The Rydberg–Ritz combination principle

Another early empirical result of spectroscopy was the *Rydberg–Ritz combination principle*, which stated that the values of the reciprocal wavelengths of many spectral lines for a given element could be 'predicted' by computing sums of the reciprocal wavelengths of other lines, taken in pairs; see Ritz (1908) for a treatment in English.

This can be seen to be a very direct consequence of the Balmer–Bohr formula. For a given element X with some Rydberg constant R_X, equation (1.7) indicates that the transition wavelengths behave as

$$\frac{1}{\lambda_{i\to f}} = R_X\left(\frac{1}{n_f^2} - \frac{1}{n_i^2}\right).$$

Suppose that there is an intermediate energy level k, between i and f. Then for the separate transitions $i \to k$ and $k \to f$, we would have

$$\frac{1}{\lambda_{i\to k}} = R_X\left(\frac{1}{n_k^2} - \frac{1}{n_i^2}\right).$$

$$\frac{1}{\lambda_{k\to f}} = R_X\left(\frac{1}{n_f^2} - \frac{1}{n_k^2}\right).$$

Adding these gives

$$\frac{1}{\lambda_{i\to k}} + \frac{1}{\lambda_{k\to f}} = R_X\left(\frac{1}{n_k^2} - \frac{1}{n_i^2}\right) + R_X\left(\frac{1}{n_f^2} - \frac{1}{n_k^2}\right) = R_X\left(\frac{1}{n_f^2} - \frac{1}{n_i^2}\right).$$

The result is exactly $1/\lambda_{i\to f}$. From an energy perspective, this is not surprising: the sum of the energy drops $i \to k$ and $k \to f$ must be equal to that of executing the $i \to f$ transition directly. But prior to the advent of the Bohr model, this combination principle was just another mystery of spectral bookkeeping.

2.6 Angular momentum

A key aspect of the Bohr model involves the magnitude of the angular momentum of the electron/nucleus system. Recall from basic mechanics that angular momentum is formally a vector quantity, $\boldsymbol{L} = \boldsymbol{r} \times \boldsymbol{p}$, where $\boldsymbol{r}$ is position and $\boldsymbol{p}$ linear momentum: $\boldsymbol{p} = m\mathbf{v}$. For a mass m in a circular orbit of radius r at speed v, the magnitude of L is given by $L = mvr$ because $\boldsymbol{r}$ and $\boldsymbol{p}$ are always perpendicular to each other.

Accounting for the motions of both the electron and nucleus then gives

$$L = m_e v_e r_e + m_{\text{nuc}} v_{\text{nuc}} r_{\text{nuc}}.$$

Replace v_{nuc} and r_{nuc} with equations (1.8) and (1.9):

$$L = m_e v_e r_e\left[\frac{m_e + m_{\text{nuc}}}{m_{\text{nuc}}}\right] = \frac{m_e^2 v_e r_e}{\mu}, \tag{2.19}$$

where μ is the reduced mass of equation (1.11). Now replace v_e and r_e with equations (2.4) and (2.6):

$$L = \frac{m_e^2}{\mu}\left(\frac{Ze^2\mu}{2\varepsilon_o m_e nh}\right)\left(\frac{\varepsilon_o h^2}{\pi Z m_e e^2}\right)n^2 = \left(\frac{h}{2\pi}\right)n. \tag{2.20}$$

The conclusion here is that, quite independent of the masses of the electron and nucleus, L is quantized as integer multiples of $h/2\pi$. A corollary conclusion is that if the electron changes its orbital state, the angular momentum of the system must change. But if no external torque acts on the system, angular momentum must be conserved, so we must also conclude that photons possess quantized angular momentum in the amount of integer multiples of $h/2\pi$. This angular momentum is now called the 'spin' of the photon, a term that should not be interpreted too literally. Photons are now considered to be members of the class of elementary particles known as *Bosons*, all of which possess spins of integer multiples of $h/2\pi$. Bosons are named after Indian physicist Satyendra Nath Bose (1894–1974). All elementary particles that are not Bosons are known as *Fermions*, after Enrico Fermi (1901–54); all Fermions possess spins that are half-integer multiples of $h/2\pi$.

$h/2\pi$ is given its own name and symbol: the 'reduced' Planck's constant, $\hbar$, read as 'h-bar'. The term 'reduced' is used here in a different sense than as in reduced mass. There are many 'reduced' quantities in physics, all defined as some quantity divided by 2π, and designated with the same symbol as the quantity in question but with a stroke through it. For example, one sometimes encounters reduced wavelength, $\lambda\!\!\!^{-} = \lambda/2\pi$.

Many textbooks begin the derivation of the Bohr model by assuming that angular momentum is quantized in this way, and applying the rule to the orbit of the electron: $m_e v_e r_e = nh/2\pi$. Then one solves for v_e, and back-substitutes into the centripetal force expression of equation (1.14), and solves for the radius r: the result is exactly the permitted orbits in terms of the Bohr radius. The result for r goes back into the total energy of equation (1.18), and the permissible energies emerge. Bohr did remark in the first paper of his trilogy that his model implied that angular momentum would be quantized in this way. (To see that the results are the same even if the reduced mass is accounted for, use equation (2.19) to set $L = m_e^2 v_e r_e/\mu = n(h/2\pi)$, and then follow the above recipe. Do it as an exercise.)

Angular momentum is also quantized in the wave-mechanical description of the hydrogen atom, albeit in a different way than in the Bohr model. One of the differences is that in the wave picture, the ground state ($n = 1$) of hydrogen, notated as the *1s* state, has zero angular momentum: the electron occupies a probability 'cloud' which is spherically symmetric about the nucleus; see section 3.8. In the Bohr model, the ground state has angular momentum $L = \hbar$. The energy levels of hydrogen as determined by the Bohr and wave theories are identical, but in the wave theory there is no concept corresponding to an electron 'orbit'; this is discussed briefly in chapter 3.

Exercise

As another (very unrealistic) model of an atom, consider a solid sphere of mass M and radius R rotating about an axis through its center in accordance with the Bohr quantization condition on angular momentum, $L = n(h/2\pi)$. Show that the possible rotational energies are given by $E_n = 5n^2h^2/16\pi^2MR^2$. From this result, derive an expression for the wavelength of a photon emitted as this system undergoes a transition from state n +1 down to state n. Evaluate your answer numerically for the case of M equal to the mass of a proton, $R = 1$ Å, and $n = 2$. Also, what value would n have to be to give wavelengths in the optical region of the spectrum, $\lambda \sim 5000$ Å, for an $n + 1$ to n transition?

HINTS: the angular momentum of a spinning solid object is $L = I\omega$, where I is its moment of inertia and ω its angular rotation speed in radians per second. The rotational kinetic energy is $I\omega^2/2$; there is no potential energy to consider in this problem.

[Ans: $\lambda_{n+1\to n} = 16\,\pi^2cMR^2/5h\,(2n + 1)$. $\lambda_{3\to 2} \sim 478\,000$ Å; $n_{\text{optical}} \sim 240$]

2.7 Unpacking paper I: the correspondence principle

With the fundamentals of the Bohr theory now established, we backtrack to examine more closely how Bohr set them out. While this may seem redundant, it will be valuable: it never hurts to recap a lesson to remind you what you have learned, and it can be illuminating to see how Bohr struggled with the revolutionary aspects of his new model.

Paper I runs to 24 pages, divided into an Introduction and five sections. The Introduction, which fills two pages, reminds readers of Ernest Rutherford's nuclear atom-model as a successor to Thomson's, and comments on the longstanding problem of the stability of atoms against their predicted prompt collapse. Planck's concept of radiation quanta is introduced, with particular mention of how it had proven valuable for explaining the photoelectric effect. This sets the stage for the observation that there '... seems to be a general acknowledgment of the inadequacy of the classical dynamics in describing the behavior of systems of atomic size', which is followed by the assessment that 'Whatever the alteration in the laws of motion of the electrons may be, it seems necessary to introduce in the laws in question a quantity foreign to the classical electrodynamics, i.e., Planck's constant .... By the introduction of this quantity the question of the stable configuration of the electrons in the atoms is essentially changed, as this constant is of such dimensions and magnitude that it, together with the mass and charge of the particles, can determine a length of the order of magnitude required'. This length is the Bohr radius.

Section 1 of the paper gets directly to the idea of electrons binding to positive nuclei. Bohr first reminds readers of Planck's $E = nhf$ formula (he used τ in place of our modern n), posits the assumption that an electron which binds to a nucleus from an initially great distance sheds a quantum of frequency one-half of its eventual orbital frequency, and then presents the formulae for the energies and radii of the permitted 'stationary states', presuming that the mass of the electron can be neglected in comparison with that of the nucleus. The Bohr radius and ionization potential of hydrogen are presented, and are shown to be in satisfying agreement with the known sizes of atoms and measured potentials. All of this emerges on page 5 of the paper.

A theme that recurs throughout the paper is exploring how the new theory connects to the predictions of classical electrodynamics. Recall that in that theory, a charged particle vibrating at some frequency f will emit electromagnetic radiation of the same frequency; Bohr termed this a 'mechanical' frequency. In contrast, the frequencies of quanta in his analysis arise from the differences in energies between his stationary states. The first detailed appearance of this issue is on pages 6 and 7 of the paper, where Bohr dissects Nicholson's attempt to explain stellar spectral features with atomic models involving rings of orbiting electrons, which are taken to emit quanta in accordance with the frequencies of presumed vibrational modes in which they are moving. However, Bohr points out that this picture must be fundamentally flawed on the rationale that as soon as an electron emits any energy, its vibrational frequency will change, and so there is no way for the model to reproduce the consistently-seen spectrum of, for example, the Balmer lines in hydrogen. There will be more to say about his 'mechanical' frequencies in connection with what would become known as the 'Correspondence Principle'; this is discussed below.

Section 2 of the paper begins on page 8, and addresses the question of line spectra. Bohr now makes the assumption that transitions between energy states are accompanied by the emission of quanta whose energies are equal to the difference in energies between the two states involved, and again invokes Planck to write this in frequency form via equation (1.1). The Balmer formula then makes its appearance, with the inference that it corresponds to $n_{\text{final}} = 2$. It is also remarked that $n_{\text{final}} = 3$ corresponds to the then-known Paschen series, and it is predicted that other series with $n_{\text{final}} = 1, 4, 5, \ldots$ await discovery. Bohr then recovers the Rydberg constant, posits that the Pickering–Fowler observations described in section 2.4 are due to helium, and points out that the theory also explains the Rydberg–Ritz combination principle of section 2.5. It is quite remarkable what Bohr squeezed into the four pages of this section.

Section 3 of the paper returns to examine some of the assumptions made, showing that the correct Balmer formula will arise only if the one-half-of-the-final-orbital-frequency assumption is adopted. It is also in this section that the Correspondence Principle makes its first appearance. Bohr is still concerned with connecting his new theory to the mechanical frequencies of classical electrodynamics, and approaches this issue in a limiting-case way as follows.

Consider an electron in orbit n, with $m_p \gg m_e$, so that the reduced mass $\mu \sim m_e$. Then from equations (2.4) and (2.6), the orbital frequency f_n of the electron, $v_e/2\pi r_e$ (i.e., the number of orbits executed per second), will be ($Z = 1$)

$$f_n = \left(\frac{m_e e^4}{4\varepsilon_o^2 h^3}\right)\frac{1}{n^3}.$$

Now imagine that the electron transits down by one orbit, to orbit $n - 1$. From equation (2.14), the frequency of the emitted photon, $f^{\text{quant}}_{n \to n-1}$, will be ($\mu = m_e$)

$$f^{\text{quant}}_{n \to n-1} = \frac{c}{\lambda_{n \to n-1}} = \left(\frac{m_e e^4}{8\varepsilon_o^2 h^3}\right)\left(\frac{1}{(n-1)^2} - \frac{1}{n^2}\right),$$

which reduces to

$$f^{\text{quant}}_{n \to n-1} = \left(\frac{m_e e^4}{8\varepsilon_o^2 h^3}\right)\left[\frac{(2n-1)}{n^2(n-1)^2}\right]. \tag{2.21}$$

Note the notational nuance here: the 'quantal' frequency of the photon has been adorned with the superscript 'quant', a reminder that it is a very different type of frequency than the mechanical one.

Now suppose that $n \gg 1$. The square bracket in equation (2.21) reduces to $\sim 2/n^3$, leaving

$$f^{\text{quant}}_{n \to n-1} \sim \left(\frac{m_e e^4}{4\varepsilon_o^2 h^3}\right)\frac{1}{n^3} \quad (n \gg 1).$$

This result is *identical* to that for the mechanical frequency f_n. It occurred to Bohr that if n is large, then the atom will become increasingly 'macroscopic' in size, where one might expect classical physics to prevail. The calculation supports this intuition in that it tells us that for large values of n, the frequency of the photon emitted in the $n \to n - 1$ transition will be essentially identical to the classical 'mechanical' orbital frequency of the electron in its initial (or final) orbit. This identity comprised what Bohr saw as a 'correspondence' between classical electrodynamics and his new quantum atomic dynamics. In subsequent attempts to elaborate the quantal theory to accommodate multi-electron atoms and molecules, and later during the development of formal quantum mechanics during the 1920s, this Correspondence Principle became a touchstone for Bohr and his collaborators. There is no formal mathematical definition of the Correspondence Principle; rather, it is expressed qualitatively with statements along the lines of 'The predictions of quantum mechanics should concur with those of classical theory in domains in which the latter are known to be valid, typically situations involving high quantum numbers where the effects of quantization become less and less important'. For more advanced students, some applications of the Correspondence Principle can be found in Crawford (1989).

Continuing with our tour through Paper I, angular momentum makes its first appearance on page 15 (still within section 3), along with the realization that the theory could be developed based on the assumption that ' … the angular momentum of the electron round the nucleus … is equal to an entire multiple of a universal value …', the universal value being $h/2\pi$. Bohr generously gives Nicholson credit for pointing out the possible importance of angular momentum in atomic systems.

Section 4 of the paper points out that the theory can explain Kirchhoff's laws by imagining that an atom can absorb energy from the outside, causing an electron to raise its energy level and thus explain why absorption spectra are the reverse of emission spectra; this section also offers further reflections on the mechanical aspects of atomic systems, and suggests that the emission/absorption theory is somewhat akin to Einstein's treatment of the photoelectric effect.

In his last section, Bohr deals briefly with treating multi-electron atoms as comprising electrons distributed in rings orbiting the nucleus, but comes to the conclusion that there is no convincing way to explain line spectra on this basis. At the very end of the paper, he reiterates that angular momentum has assumed a primary role, summarizing his fundamental working hypothesis of atomic structure as 'In any molecular system consisting of positive nuclei and electrons in which the nuclei are at rest with respect to each other and the electrons move in circular orbits, the angular momentum of every electron round the center of its orbit will in the permanent state (that is, $n = 1$) of the system be equal to $h/2\pi$, where h is Planck's constant'. As described in section 2.6, this assumption, which we now know to be untrue, became the starting point for many textbook derivations of Bohr's theory. An alternate derivation based on assuming this and an energy-minimization argument appears in section 3.7.

In the end, the model of hydrogen advanced in Bohr's Paper I was a strange hybrid of classical and quantum (Planck's constant) concepts linked by the Correspondence Principle and glued together by a hodgepodge of unusual assumptions: electrons can radiate when they are binding to an electron or transiting, but not when ensconced in a stationary state; and angular momentum, one of the most venerable quantities of classical mechanics, seems to be intimately related to Planck's constant, which arose in the study of thermodynamics. But strange as it was, the successes of the approach could not be denied.

2.8 Initial reactions to the theory

It is rare for a new scientific theory to be accepted without dissent, and Bohr's atomic model, with its strange stationary states and quantum-jumping assumptions, was no exception. Some of the leading atomic structure researchers of the day simply ignored the theory, while others, notably John Nicholson, argued against it, and yet others treated it with faint praise. The distinguished Cambridge mathematician and astronomer James Jeans remarked that 'The only justification at present put forward for these assumptions is the very weighty one of success'.

One obvious objection was raised by Rutherford when Bohr sent him a draft of his first paper in March, 1913. In a letter to Bohr, Rutherford remarked that 'There appears to me one grave difficulty in your hypothesis, … how does an electron decide what frequency it is going to vibrate at when it passes from one stationary state to the other? It seems to me that you would have to assume that the electron knows beforehand where it is going to stop'. Bohr had no real answer for this question, which would not be resolved until the development of quantum mechanics proper provided a way of computing the intensities of spectral lines. New theories are not born whole.

The main point of objection in many quarters was that Bohr's theory departed so radically from classical physics, most notably in its assumption that spectral frequencies reflected transitions between energy levels as opposed to the notion that the electron should always emit at a frequency identical to that of a vibrational or orbital motion, although the classical view would still be plagued by the classical collapse problem. This would not, however, be the first time in the history of physics that radical new ideas seemed to be needed to explain phenomena: the two outstanding examples from that era are Planck's introduction of his energy/frequency relationship to understand blackbody spectra, and Einstein's postulate of the universality of the speed of light in response to failures to detect the predicted motion of the Earth through a universe-filling 'ether'. It is not uncommon for new theories to take a generation to become established as proponents of superseded ones die off.

As described by historian of physics Helge Kragh, rival theories held considerable credibility through about 1915 in Britain in particular; much of the discussion in this section is drawn from this paper (Kragh 2011). While Thomson, the master atom-builder, eventually abandoned his own plum-pudding model with the advent of the Rutherford atom, he regarded Bohr's model as a mathematical construct without much physical basis and essentially ignored it, instead developing a new model of his own in 1913 which required positing an inverse-cube repulsive force acting on electrons within atoms in addition to the usual inverse-square attractive Coulomb force, but with the latter now confined to act within radially-oriented tubes. With these and some other *ad hoc* assumptions he did succeed in developing an explanation of some x-ray data, but this model was in conspicuous conflict with alpha-scattering data. Bohr and Thomson, not on the best of terms since they had first met, largely ignored each other.

Others developed extrapolations of the original plum-pudding model to incorporate rotation and vibrations contrived to give the desired frequencies, or postulated complicated schemes of negatively charged shells surrounding a positive nucleus. Bohr could afford to ignore many of these concepts as their proponents often lay outside the mainstream of physics (mathematicians and engineers, for example), but most serious were the objections of John Nicholson. A central tenet of much of Nicholson's work was the idea that the 'primary' forms of elements did not exist terrestrially, but only in stars and nebulae. Spectroscopic data led him to postulate elements with names such as 'coronium' (detected in spectra of the Sun's corona, and which he took to be the simplest element, with two electrons orbiting a positive nucleus), 'nebulium', 'protofluorine', 'protohydrogen', and 'archonium' in addition to hydrogen. An underlying point of confusion here was that it was not yet clear if it was the atomic number or the atomic weight of an element which dictated its place in the periodic table; this was really only sorted out with the x-ray work of British physicist Henry Moseley in 1914, which is described in section 3.3. Mosley's work would significantly bolster acceptance of Bohr's theory in comparison to its rivals.

As Kragh describes it, Nicholson's work did have some credible elements (bad pun intended), such as adopting Planck's constant as a fundamental constant of nature and emphasizing the role of angular momentum, but it also had an underlying air of numerology. Bohr and Nicholson debated the merits of their theories in a series of back-and-forth articles and letters in *Nature* and *The Philosophical Magazine*, but after about a year Bohr apparently decided that the dispute was not worth the trouble. After 1915, Nicholson stopped criticizing Bohr, at least in public. His 'primary elements' would prove to be ordinary elements rendered into in high states of ionization (e.g., 13-times ionized iron) in the exotic conditions of high temperature and low pressure that exist in the outer layers of stars. In the meantime, other researchers began to extend Bohr's theory into new territory. In particular, German physicist Arnold Sommerfeld began putting Bohr's theory on a more rigorous footing by incorporating Einstein's theory of relativity and what was known as Hamilton–Jacobi theory, an advanced formulation of classical mechanics. A motivation for this was to develop an explanation of the so-called fine structure of spectral lines; this is described briefly in section 3.8.

The speculations of Thomson, Nicholson, and others may seem amusing curiosities to us now, but beware of the smugness of judging the past against current knowledge. All theories evolve in response to knowledge brought forth by new experimental technologies and new mathematical techniques. In their time, these were serious theories advanced by established, knowledgeable researchers. A century from now, many of our current ideas may be regarded as head-shakers. What is commendable about Bohr is that when new and improved ideas emerged, he actively engaged them and helped develop and interpret them, ensuring a long and fruitful career.

References

Beyer A *et al* 2017 *Science* **358** 79–85

Bohr N 1913a *Philos. Mag. Ser. 6* **26** 1–24

Bohr N 1913b *Philos. Mag. Ser. 6* **26** 476–502

Bohr N 1913c *Philos. Mag. Ser. 6* **26** 857–75

Bohr N 1913d *Nature* **92** 231–2

Brackett F S 1922 *Nature* **109** 209

Crawford F S 1989 *Am. J. Phys.* **57** 621–8

Ellison F O and Hollingsworth C A 1976 *J. Chem. Educ.* **53** 767–70

French A P and Kennedy P J (ed) 1985 *Niels Bohr: A Centenary Volume* (Cambridge, MA: Harvard University Press)

Griffiths D J 1995 *Introduction to Quantum Mechanics* (Upper Saddle River, NJ: Prentice-Hall)

Höglund B and Mezger P 1965 *Science* **150** 339–40 347–8

Humphreys C J 1953 *J. Res. Natl. Bur. Stand.* **50** 1–6

Kragh H 2011 *Phys. Perspect.* **13** 4–35

Lyman T 1914 *Phys. Rev.* **3** 504

Nagaoka H 1904 *Philos. Mag. Ser. 6* **7** 445–55

Nicholson J W 1911 *Mon. Not. R. Astron. Soc.* **72** 49–64

Pais A 1991 *Niels Bohr's Times, in Physics, Philosophy, and Polity* (Oxford: Oxford University Press)

Paschen F 1908 *Ann. Phys. Lpz.* **332** 537–70

Pfund A H 1924 *J. Opt. Sci. Am.* **9** 193–6

Ritz W 1908 *Astrophys. J.* **28** 237–43

IOP Publishing

The Bohr Atom
A guide
Bruce Cameron Reed

Chapter 3

The Bohr model II: corollaries

This chapter examines several corollary issues associated with the Bohr model. These include graphical depictions of the theory via orbital and energy-level diagrams; the fine-structure constant; the success of the theory in explaining x-ray data; why gravity is unimportant in the Bohr model; two alternate derivations of the theory; how Bohr attempted to extend the theory to consider multi-electron atoms and molecules; how other researchers incorporated relativistic effects into the theory, and a brief description of current quantum-mechanical concepts of the structure of hydrogen 'electron clouds'.

3.1 Orbit and energy-level diagrams

Various methods for pictorially depicting the Bohr–Balmer formula have been developed. These tend to be either schematic or precisely technical; one of each type is described in this section.

One method of depicting the theory would be to extend the wavelength scale in figure 1.5 to run from zero to infinity so that *all* spectral series could be displayed, but this has obvious problems of practicality. Instead, a common textbook illustration is of the format of figure 3.1, where transitions between the first few orbits are shown as arrowed lines, with attached labels depicting wavelengths. This method has the advantage of reinforcing the concept of permitted orbits, but suffers from the limitation that it is difficult to draw the orbits to scale since they space as n^2.

More technically precise—and able to fit on a single page—is the notion of an *energy-level diagram*, based on equations (2.7) and (2.8):

$$E_n = -(13.598\ \text{eV})\frac{1}{n^2}.$$

This is shown in figure 3.2 for hydrogen. Here, energy is plotted increasing upward on the vertical axis; the horizontal axis has no physical meaning, and is used only to space out the various spectral series for clarity. Such plots are also known as

doi:10.1088/978-0-7503-3611-6ch3

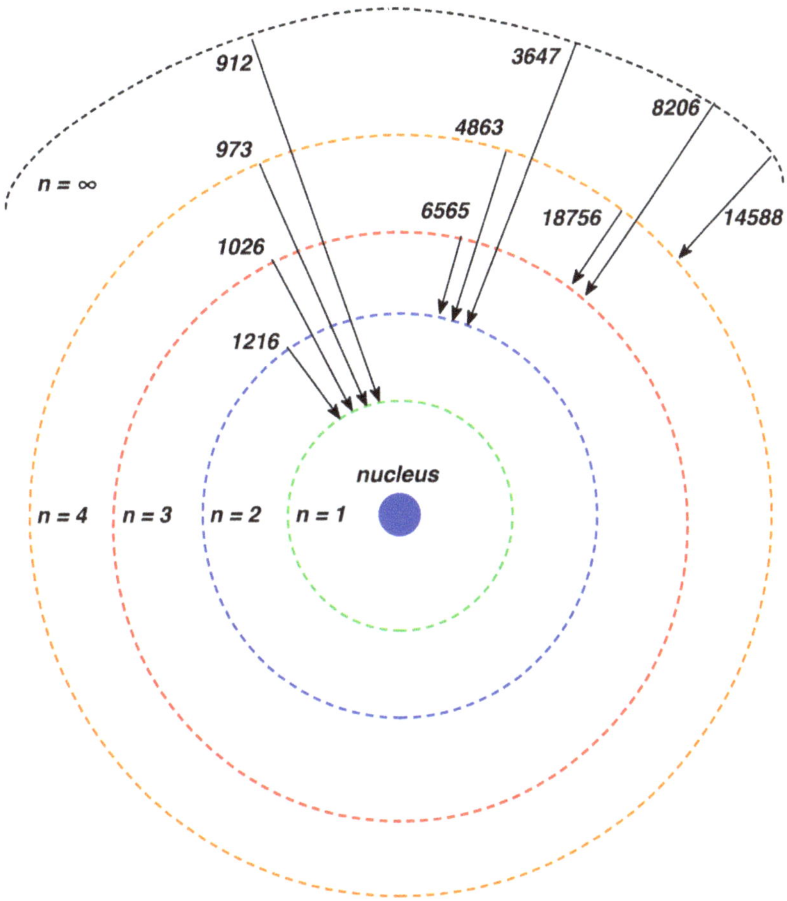

Figure 3.1. Schematic illustration of emission transitions between the $n = 1$ and $n = 4$ levels of hydrogen, and the series-limit transitions originating at $n_{\mathrm{initial}} = \infty$. Each transition is labeled with its wavelength in Ångstroms, rounded off to the nearest whole number. The orbital radii are not to scale; colors are for illustrative purposes only.

Grotrian diagrams, after German astronomer Walter Grotrian (1890–1954), who first developed them in 1928. Given energy levels run horizontally across the diagram, and, as with the orbital diagram above, arrowed lines indicate possible transitions. All energies are negative, and approach zero from below as $n \to \infty$; zero energy is at the top of the diagram. The advantage of this scheme is that if the levels are drawn to scale on the energy axis (they are not in figure 3.2), the lengths of the transition arrows are directly proportional to their energies, and hence corresponding frequencies via $E = hf$. The disadvantage of this scheme is that the energy levels become increasingly crowded together as n increases. Electrons that are not initially bound to the nucleus, that is, free electrons with $E > 0$, can become bound by releasing just enough energy to bring them into one of the permitted energy levels; there is said to be a *continuum* of energy levels with $E > 0$. An advantage of this

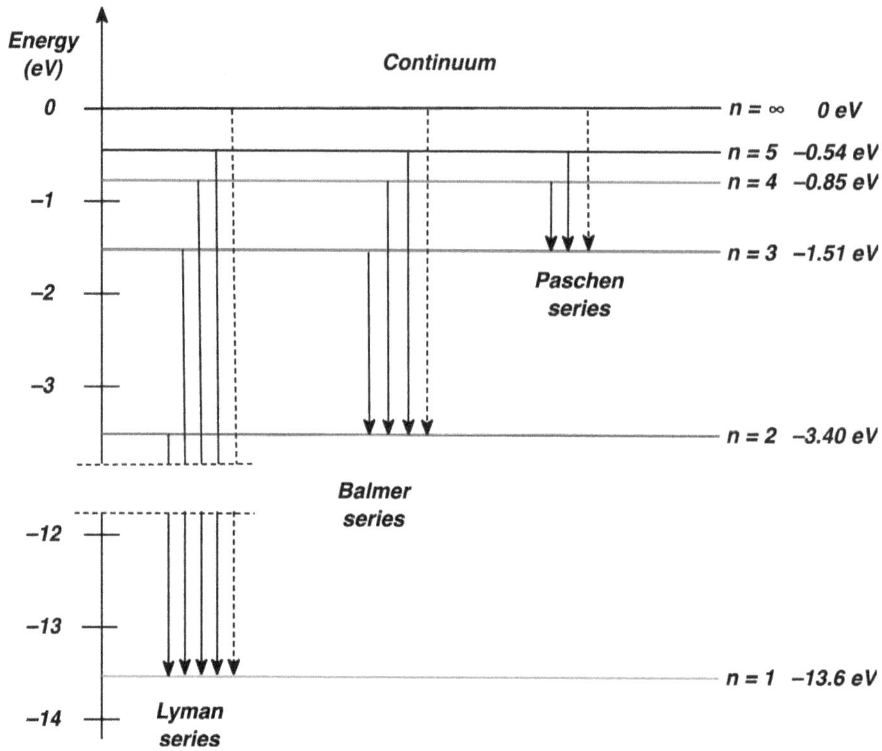

Figure 3.2. Grotrian diagram for hydrogen. For each series shown, the series-limit transition is drawn with a dashed line. Not to scale. The colors of the levels are not meaningful, but are the same as the corresponding orbits in figure 3.1.

approach is that the horizontal lines can be 'forked' to depict the fine-structure of spectral lines, as described in section 3.8.

3.2 The fine-structure constant

From equation (2.4), the speed of the electron in orbit n is given by

$$v_e = \frac{Ze^2\mu}{2\varepsilon_o m_e n h}. \tag{3.1}$$

Consider again the case that the proton is much more massive than the electron, so $\mu \sim m_e$. To create a dimensionless quantity here, form the ratio of the electron speed to the speed of light, and take $Z = 1$ and $n = 1$:

$$\frac{v_e}{c} = \frac{e^2}{2\varepsilon_o hc}.$$

This ratio is known as the 'fine-structure constant', and is usually designated by the symbol α. It evaluates to

$$\alpha = \frac{e^2}{2\varepsilon_o hc} \sim 7.29735 \times 10^{-3} \sim \frac{1}{137.04}. \tag{3.2}$$

This result tells us that the electron's speed in the $n = 1$ Bohr orbit is somewhat less than 1% of the speed of light, so a non-relativistic derivation of the Bohr model is justifiable. Also, if we stick with a non-relativistic calculation, this result shows that we must have $Z < 137$ to keep $v/c < 1$. Surprisingly, there are about 90 naturally-occurring elements.

The fine-structure constant was introduced by German theoretical physicist Arnold Sommerfeld in 1916 in an extension of the Bohr model to include elliptical orbits and relativistic effects in an effort to analyze the fine structure of spectral lines of hydrogen; see section 3.8. α also enters the more advanced field of quantum electrodynamics in quantifying the strength of interactions between electrons and photons.

3.3 X-rays

Beyond its remarkable success in explaining the Balmer formula and the Rydberg constant, the first independent success of the Bohr model came in explaining a regularity in the x-ray spectra of metals. This success significantly bolstered acceptance of Bohr's theory.

Since the discovery of x-rays by Wilhelm Röntgen in late 1895, it had been known that the means to produce them was to fire highly-accelerated electrons at a metal target within an evacuated glass bulb. The vacuum prevented x-rays from being slowed and deflected; the struck target ('anode') would emit x-rays which could penetrate through the bulb, pass through a subject, and be recorded on a photographic plate. But precisely what was the origin of this high-energy form of light remained a mystery until publication of the Bohr model.

A key breakthrough was made by British physicist Henry Moseley in two papers published about the time that the Bohr model appeared (Moseley 1913, 1914; figure 3.3). Experimenting with over three dozen targets of different metals, Moseley found that x-rays were emitted at a variety of frequencies, but there was a regularity. This was that the highest frequency emitted by an element of atomic number Z behaved empirically as

$$f_{\max} \sim \frac{3}{4}K(Z-1)^2,$$

where the constant K was identical to that which appeared in the Balmer formula when it was rearranged to express frequencies as opposed to wavelengths. This is equation (2.14), generalized for atomic number Z and setting $\mu \to m_e$:

$$f_{i\to f} = \frac{c}{\lambda_{i\to f}} = \underbrace{\left(\frac{Z^2 e^4 m_e}{8\varepsilon_o^2 h^3}\right)}_{K}\left(\frac{1}{n_{\text{final}}^2} - \frac{1}{n_{\text{initial}}^2}\right).$$

Figure 3.3. Henry Moseley (1887–1915) with an x-ray tube, about 1910. *Source*: Public domain; https://commons.wikimedia.org/wiki/File:BigMoseleyCard.jpg.

Consider $n_{\text{initial}} = 2$ and $n_{\text{final}} = 1$, as in the Lyman series:

$$f_{2\to 1} = \frac{3}{4}KZ^2.$$

But for having $(Z - 1)$ in place of Z, this is identical to Moseley's empirical formula. Ignoring this discrepancy for the moment, creation of the highest-frequency x-rays can be interpreted as incoming electrons penetrating deep inside an atom of the metal target and knocking an electron out of the $n = 1$ orbit, whose place is then taken by an electron transiting down from the $n = 2$ orbit. The $(Z - 1)$ instead of Z was later 'understood' in the context of Bohr's 'shell' model of electron orbits (section 3.8): the innermost shell can contain two electrons, so the one remaining after the knock-out provides some 'shielding' against the charge of the nucleus as seen by more distant electrons.

X-ray photons are of very high energy, and it is for this reason that your dentist has you wear a lead apron, even though the beams are very focused. From equation (2.14), taking the reduced mass to be m_e and modifying the Rydberg constant for general atomic number Z:

$$\lambda_{i\to f} = R_\infty^{-1}\left(\frac{1}{n_{\text{final}}^2} - \frac{1}{n_{\text{initial}}^2}\right)^{-1} = \frac{(911.3\ \text{Å})}{Z^2}\left(\frac{1}{n_{\text{final}}^2} - \frac{1}{n_{\text{initial}}^2}\right)^{-1}.$$

For a titanium anode with $Z = 22$ and $(n_{\text{initial}}, n_{\text{final}}) = (2, 1)$:

$$\lambda = \frac{4}{3}\frac{(911.3)}{484}\text{Å} = 2.51\ \text{Å}.$$

From equation (1.3), the energy is

$$E = \frac{12\,400}{\lambda_{\text{Å}}}\text{eV} \sim 4.9\ \text{kV}.$$

Mosley's work helped to sort out atomic numbers for some elements and place them in their proper locations in the periodic table. His work was considered so fundamental in early atomic physics that it has been speculated that he would have been awarded a Nobel Prize had he not been killed in action while serving with the Royal Engineers of the British Army at the Battle of Gallipoli on August 10, 1915, during World War I.

Further support for Bohr's theory emerged from a completely different direction almost simultaneously with Moseley's work in April, 1914, when an experiment performed by James Franck and Gustav Hertz revealed direct evidence for quantized energy levels of electrons in atoms. Franck and Hertz thought they were measuring ionization potentials, and were unaware of Bohr's theory.

3.4 Overlapping and non-overlapping spectral series

It was commented in section 3.4 that only the Lyman and Balmer series of hydrogen are completely unentangled from neighboring sequences. A proof of that assertion is developed here.

A series of spectral lines corresponds to all possible transitions from higher-energy orbits to the same lower-energy orbit. From the Balmer–Bohr formula [equation (2.14)] for wavelengths of a given series 's' where $n_{\text{final}} = n_s$ and $n_{\text{initial}} = n$ in general, we have

$$\lambda = R_{\text{H}}^{-1}\left[\frac{1}{n_s^2} - \frac{1}{n^2}\right]^{-1} \quad (n > n_s).$$

For a fixed value of n_s, the lowest and highest wavelengths possible correspond to $n = \infty$ and $n = n_s + 1$, respectively:

$$\lambda_{\text{low}}(n_s) = R_{\text{H}}^{-1}\, n_s^2$$

and

$$\lambda_{\text{high}}(n_s) = R_{\text{H}}^{-1}\left[\frac{n_s^2(n_s + 1)^2}{2n_s + 1}\right].$$

As n_s increases, the wavelength of the lowest-wavelength line for a series becomes progressively greater. To determine if a series of lines will be isolated from neighboring series is equivalent to demanding that its lowest wavelength line lies at a wavelength greater than the highest wavelength possible for lines of series $n_s - 1$, and also that its highest-wavelength line lies below the lowest-wavelength line for series $n_s + 1$, that is,

$$\left.\begin{aligned} \lambda_{\text{low}}(n_s) &\geqslant \lambda_{\text{high}}(n_s - 1) \\ \lambda_{\text{high}}(n_s) &\leqslant \lambda_{\text{low}}(n_s + 1) \end{aligned}\right\}.$$

Consider these independently:

$$\lambda_{\text{low}}(n_s) \geqslant \lambda_{\text{high}}(n_s - 1) \Rightarrow n_s^2 \geqslant \left[\frac{(n_s - 1)^2 n_s^2}{2(n_s - 1) + 1}\right] \Rightarrow n_s^2 - 4n_s + 2 \leqslant 0.$$

Only $n_s = 1$, 2, and 3 satisfy this requirement. The second constraint reduces to

$$\lambda_{\text{high}}(n_s) \leqslant \lambda_{\text{low}}(n_s + 1) \Rightarrow \left[\frac{n_s^2(n_s + 1)^2}{2n_s + 1}\right] \leqslant (n_s + 1)^2 \Rightarrow n_s^2 - 2n_s - 1 \leqslant 0.$$

Only $n_s = 1$ or 2 satisfy this constraint, meaning that only $n_s = 1$ or 2 satisfies both constraints: as claimed, only the Lyman and Balmer series are disentangled from other series. Further, since the Rydberg constant canceled in this calculation, this conclusion holds for *any* one-electron atom, at least so far as we can trust a non-relativistic calculation.

3.5 Whither gravity?

The entire discussion so far has neglected any effects of gravity. Electrons and protons do have mass, so they will exert gravitational forces on each other. Might this alter the possible energy levels of the atom? It turns out that, for practical purposes, the answer is a definite 'no'.

Consider an electron and a proton separated by distance r. The magnitudes of the electrical and gravitational attractions between them are

$$F_{\text{elec}} = \frac{e^2}{4\pi\varepsilon_o r^2}$$

and

$$F_{\text{grav}} = \frac{Gm_e m_p}{r^2}.$$

The ratio of these is independent of r:

$$\frac{F_{\text{elec}}}{F_{\text{grav}}} = \frac{e^2}{4\pi\varepsilon_o G m_e m_p}.$$

If you put in the numbers, you will find $F_{\text{elec}}/F_{\text{grav}} \sim 2.3 \times 10^{39}$. The power of 10 is not a misprint: electricity is a vastly stronger force than gravity. It takes an entire planet to muster an appreciable gravitational force. We can neglect gravity in the Bohr atom.

3.6 The de Broglie matter-wave approach

In his 1924 PhD thesis, French physicist Louis Victor de Broglie (1892–1987; pronounced 'de broy') proposed that if light could behave like particles and carry momentum, perhaps ordinary particles could behave like waves and possess associated wavelengths.

Einstein's 1905 theory of relativity showed that light of wavelength λ carries momentum p according as

$$p = \frac{h}{\lambda},$$

where h is, as usual, Planck's constant. In classical non-relativistic mechanics, momentum is the product of mass and velocity, mv. Applied to an electron, de Broglie's hypothesis then predicts a wavelength according as

$$p = \frac{h}{\lambda} = m_e v_e \Rightarrow \lambda^{\mathrm{deB}} = \frac{h}{m_e v_e}. \tag{3.3}$$

I attach a superscript deB to the λ here as a reminder that this λ is not an 'ordinary' wavelength in the sense of a photon wavelength. In most circumstances, de Broglie wavelengths are extremely small. For an electron moving at a speed corresponding to a single electronvolt of kinetic energy, $\lambda^{\mathrm{deB}} \sim 12$ Å, on the order of the size of an individual multi-atom molecule.

de Broglie's hypothesis, strange as it may seem, received experimental verification within three years, when Clinton Davisson and Lester Germer, working at the Bell Telephone laboratories in the United States, detected a diffraction pattern formed by electrons scattering from crystals of nickel (Davisson and Germer 1927). The next year, such patterns were detected in a purpose-built experiment by British physicist George P Thomson (JJ's son) wherein electrons were scattered upon passing through thin metallic foils. For our purposes, however, what is of interest is how the matter–wave hypothesis can be used to develop an alternate derivation of the Bohr energy levels. This proceeds by hypothesizing that permitted electron orbits must have wavelengths such that an integral number of whole wavelengths fit around the circumference of the orbit, $2\pi r$. This requires (neglecting any motion of the nucleus)

$$2\pi r_e = n\lambda^{\mathrm{deB}} = \frac{nh}{m_e v_e}.$$

Rearranging this gives

$$m_e v_e r_e = n\left(\frac{h}{2\pi}\right).$$

This is precisely the quantization of angular momentum condition that emerged from the Bohr model, from which the usual permitted orbital radii and energies can be derived. Figure 3.4 shows a conceptual sketch of an integer number of wavelengths fitting around the circumference of a circle.

De Broglie's integer-number-of-whole-wavelengths hypothesis is reminiscent of the classic laboratory experiment where a vibrating string is fixed at both ends: for constructive interference, the frequency of vibration must lead to a wavelength such that an integral number of half-wavelengths fit between the mounting posts. If you are familiar with more advanced quantum mechanics, a similar situation arises in the case of a so-called infinite potential well, where the permitted energy states are such that a half-integer number of de Broglie wavelengths fit across the width of the well.

A common error is to confuse de Broglie wavelengths—a particle property—with photon transition wavelengths. They can be related (as will be done in what follows), but doing so has no apparent physical interpretation. To make this connection, first

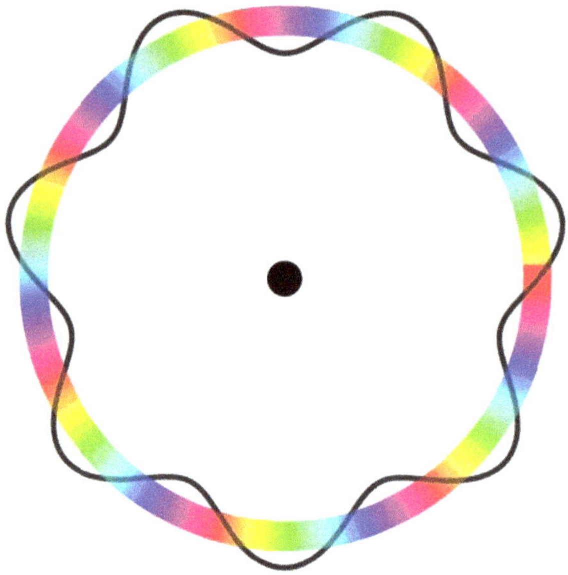

Figure 3.4. Conceptual sketch of the de Broglie representation of the $n = 7$ state of a hydrogen atom. The dot in the center represents the proton. Colors are illustrative, to designate the phase of the wave. *Source*: Peter Kuiper https://commons.wikimedia.org/wiki/File:Atom_deBroglie.png.

write the de Broglie wavelength more explicitly by substituting for the electron speed v_e from equation (3.1) and setting $\mu = m_e$:

$$\lambda^{\text{deB}} = \frac{h}{m_e v_e} = \frac{h}{m_e}\left(\frac{2\varepsilon_o nh}{Ze^2}\right) = \left(\frac{2\varepsilon_o h^2}{Zm_e e^2}\right)n.$$

From this, solve for $1/n$:

$$\frac{1}{n} = \left(\frac{2\varepsilon_o h^2}{Zm_e e^2}\right)\frac{1}{\lambda^{\text{deB}}}.$$

Now go back to Bohr–Balmer formula, equation (2.14), for some initial–final transition (again with $\mu = m_e$),

$$\frac{1}{\lambda_{n_i \to n_f}} = \frac{m_e e^4}{8\,\varepsilon_o^2 h^3 c}\left(\frac{1}{n_f^2} - \frac{1}{n_i^2}\right),$$

and replace the n's with the de Broglie wavelengths:

$$\frac{1}{\lambda_{n_i \to n_f}} = \frac{Z^2 m_e e^4}{8\varepsilon_o^2 h^3 c}\left(\frac{2\varepsilon_o h^2}{Zm_e e^2}\right)^2\left(\frac{1}{\left(\lambda_f^{\text{deB}}\right)^2} - \frac{1}{\left(\lambda_i^{\text{deB}}\right)^2}\right).$$

This simplifies to

$$\frac{1}{\lambda_{n_i \to n_f}} = \left(\frac{h}{2m_e c}\right)\left(\frac{1}{\left(\lambda_f^{\text{deB}}\right)^2} - \frac{1}{\left(\lambda_i^{\text{deB}}\right)^2}\right).$$

The factor of $h/(m_e c)$ is known as the Compton wavelength of the electron, and has the value 0.024 263 Å. If you are familiar with the process of Compton scattering wherein x-rays scatter from electrons, this factor comes into computing the change in wavelength of the scattered x-rays.

Exercise

In kinetic theory, atoms in an environment at absolute temperature T are modeled as point masses of mass m traveling at speed v, and the average kinetic energy of the atoms is $mv^2/2 = 3kT/2$, where k is Boltzmann's constant. Derive an expression for the de Broglie wavelength for an atom of mass m in terms of h, k, T, and m. At room temperature and pressure (T = 300 K), hydrogen atoms typically travel many thousands of Ångstroms between collisions. Is it plausible to treat them as point masses in such circumstances? [Ans: $\lambda = h/\sqrt{3mkT} \sim 1.5$ Å.]

3.7 An angular momentum/energy minimization approach

This section presents an unusual way of deriving the Bohr model based on the quantization of angular momentum in conjunction with energy concepts. The striking thing about this approach is that it makes no appeal to force arguments, and gives an opportunity to introduce a concept you will encounter in more advanced classes.

Take the nucleus to be fixed, and the electron, mass m_e, to be circling the nucleus in an orbit of radius r with speed v. Recall that for a circular orbit, the magnitude of the angular momentum is $L = m_e v r$. If L is quantized as $n(h/2\pi)$, then

$$v = \frac{nh}{2\pi m_e r}.$$

The kinetic energy of the electron is then

$$\text{KE} = \frac{1}{2}m_e v^2 = \frac{1}{2}m_e\left(\frac{nh}{2\pi m_e r}\right)^2 = \frac{n^2 h^2}{8\pi^2 m_e r^2}. \tag{3.4}$$

The potential energy of the electron/nucleus system is, from equation (1.17) with $\mu = m_e$,

$$\text{PE} = -\frac{Ze^2}{4\pi\varepsilon_o r}. \tag{3.5}$$

Adding equations (3.4) and (3.5) gives the total energy of the system:

$$E = \frac{n^2 h^2}{8\pi^2 m_e r^2} - \frac{Ze^2}{4\pi\varepsilon_o r}. \tag{3.6}$$

Now here comes the new concept. The only variable remaining in equation (3.6) is the orbital radius r. To determine this, make the assumption that the system settles into the minimum possible energy. The value of r that minimizes the energy can be found by setting the derivative of E to zero:

$$\frac{dE}{dr} = -\frac{n^2 h^2}{4\pi^2 m_e r^3} + \frac{Ze^2}{4\pi\varepsilon_o r^2} = 0 \Rightarrow r = \left(\frac{\varepsilon_o h^2}{\pi Z m_e e^2}\right) n^2.$$

This is exactly the same result that was obtained in section 2.2 *from Bohr's binding argument; see equation (2.6).* On back-substituting this result into equation (3.5), the same energy levels emerge as well.

This energy-minimization assumption is not just a convenient artifice: speaking somewhat loosely, it is an example of a longstanding (early-1700s) principle from mechanics known as the principle of least action; you will encounter this in an advanced dynamics class. We might in a sense regard this as another correspondence between macroscopic and atomic physics, although it does require assuming quantization of angular momentum.

3.8 Papers II, III, and beyond: the periodic table, relativity, and quantum mechanics

Paper II of Bohr's trilogy came back to his quest to understand the periodic system of the elements. An important clue in this regard was that the 'periods' of the periodic table end with particularly stable elements in the last (rightmost) column of the table, and that the number of elements in each period went as $2n^2$, with the periods numbered as $n = 1, 2, 3, \ldots$. Guided by this, his approach in paper II was that multi-electron atoms with single nuclei involve electrons orbiting the nucleus in planar rings. In this scheme, element number three, lithium ($Z = 3$), consisted of an inner ring holding two electrons plus an outer ring occupied by one lone electron. Proceeding to beryllium ($Z = 4$), its additional electron would go into the second ring. The $2n^2$ gives eight elements for the second period, which ends with neon ($Z = 10$), which would have two electrons in the inner ring plus eight in the second ring. Element 11, sodium, would then begin populating the third ring.

Analytically, Bohr was guided by the assumptions that each orbiting electron had to have an amount of angular momentum equal to $h/2\pi$ and that the electrons would distribute themselves in various rings in such a way as to minimize the total energy of the system. Quantitatively, this approach gave results in fairly good agreement with measured values of ionization energies, but he could not achieve the extreme agreement of the hydrogen results of paper I. For example, he estimated the ionization potential for the first electron of helium to be 27 eV, as opposed to the measured value of 24.6 eV. (In this sense, he actually did better than he thought at the time; the first ionization energy of helium was then estimated as ~20.5 eV.)

In paper III, Bohr attempted to extend his theory to understand the structure of multi-nuclei, multi-electron molecules, with particular attention to the simplest molecule, diatomic hydrogen, H_2. Working again from precepts of angular momentum and energy minimization, he posited that such a molecule might involve two electrons orbiting around an imaginary line joining the two protons, with the electrons keeping the protons bound against their mutual repulsion; look ahead to figure 3.11. As in paper II, calculations were only in fair agreement with experimentally measured values; his estimate of the bond strength of H_2 was only about one-half of the measured value. However, an analysis of this energetics of this system does illuminate an important physical point, so Bohr's treatment of this system is reproduced in section 3.9.

A common feature of promising new scientific developments is that they are quickly taken up by other researchers and expanded well beyond their original incarnations. In the case of the Bohr atom, one of the most vigorous proponents was Arnold Sommerfeld, who appeared briefly in section 3.2. An aspect of spectra left unaddressed in Bohr's original treatment is that they exhibit what is known as *fine structure*. This means that, when viewed with a high-resolution spectrometer capable of discriminating spectral lines separated by much less than an Ångstrom (see estimate below), they prove to be split into two or more closely-spaced components. In 1887, Albert Michelson and Edward Morley found that the red line of hydrogen at ~6563 Å is actually split into two components (Michelson and Morley 1887; this is the same Michelson and Morley of the 'ether drift experiment' that you might have encountered in studying relativity); this line is now known to be split into several components. This implies that either or both of the energy levels corresponding to the transition in question must actually comprise multiple finely-spaced energy levels. In actuality, the spacing is on the order of 10^{-5} eV.

Sommerfeld sought to extend the Bohr model to incorporate fine structure, and hit on the idea of allowing elliptical orbits in addition to Bohr's circular ones for a given quantum number n. Elliptical orbits were well-known from planetary mechanics, and since both electricity and gravity are inverse-square forces, this was a logical supposition. Sommerfeld posited that a Bohr energy level n could consist of multiple possible orbits dictated by a second quantum number k which would be quantized (of course!), with $k = 1, 2, 3, \dots n$; different values of k would determine different orbital eccentricities (and hence angular momenta), with $k = n$ corresponding to a circular orbit for a fixed n. $k = 0$ was arbitrarily excluded on the grounds that it would correspond to the electron passing through the nucleus. However, if n dictates the energy of this family of orbits, how could finely-spaced energy levels arise? In Newtonian planetary orbital mechanics, it is possible to have two orbits of exactly the same energy but different angular momenta, with the latter depending on the eccentricity of the orbit, a feature known as *energy degeneracy*. However, in classical mechanics, energy was considered to be a continuous variable, so nobody was ever concerned about positing discretely-spaced energies. Speaking a little loosely, Sommerfeld found that if he took the relativistic variation of the electron mass into account as it orbits the nucleus, this would cause the energies of the possible k-enumerated orbits for a given value of n to have slightly different energies. A more rigorous analysis of this situation shows that there are also magnetic effects involved. Electrons proved to possess intrinsic angular momentum, a property which came to be called spin, although the term is somewhat misleading. For a mental picture, imagine a small ball of charge that spins around, like the Earth spinning daily on its imaginary polar axis. A spinning charge creates a current, which will experience a magnetic force from the nucleus, and this creates an energy interaction. (The flaw in this analogy is that electrons are apparently truly point particles, with no extent.) Rather than repeat Sommerfeld's analysis, which was eventually superseded by a relativistic version of Schrödinger's equation known as the Dirac equation, I will just quote the result and use it to make a few estimates.

For a nucleus of atomic number Z and infinite mass, Sommerfeld found that the energy of a given (n, k) combination is given by

$$E_{n,k} = -\left(\frac{E_1 Z^2}{n^2}\right)\left[1 + \frac{\alpha^2 Z^2}{n}\left(\frac{1}{k} - \frac{3}{4n}\right)\right] \quad (k < n). \tag{3.7}$$

In this expression, E_1 is the magnitude of the energy of the Bohr atom ground state for an infinite-mass nucleus, 13.6 eV (i.e., equation (2.8) with μ_∞ in place of μ_{H}). α is the fine-structure constant of equation (3.2), ~1/137. Notice that if the speed of light were infinite, then we would have $\alpha = 0$, and equation (3.7) would return the normal Bohr energy levels because the second term within the square brackets would vanish. With $\alpha^2 \sim 10^{-4}$, this term will be very small, and since $k \leq n$, it will always be positive, so the quantity in the square brackets will be slightly larger than unity. This makes the energy levels slightly more negative than their non-relativistic Bohr counterparts.

The relativistic corrections are small, but corresponded roughly to what was observed. Figure 3.5 sketches the Bohr and Sommerfeld energies for the $n = 1$, 2, and 3 levels, with the splitting of the levels greatly exaggerated; the figure is not to scale.

As an example, consider the $2 \rightarrow 1$ transition, which will have a splitting of $\sim 4.5 \times 10^{-5}$ eV. What will be the difference in wavelength between the two spectral lines, which would otherwise correspond to an energy difference of 10.2 eV and wavelength ~1216 Å? We can determine this with equation (1.3), writing the factor of 12 400 Å as C (it will cancel out):

$$\lambda_{\text{Å}} = \frac{C}{E}.$$

Since the energy difference is small, we can do a differential calculation; this is a handy trick for circumstances like this:

$$\frac{d\lambda_{\text{Å}}}{dE} = -\frac{C}{E^2} \Rightarrow d\lambda_{\text{Å}} = -\frac{C}{E^2}dE = -\lambda\frac{dE}{E}.$$

Rounding off slightly and dropping the negative sign so that we are dealing just with magnitudes, this gives

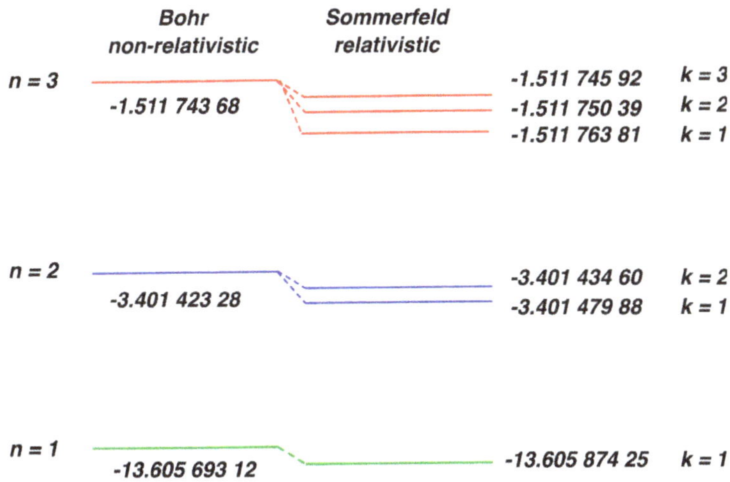

Figure 3.5. Schematic sketch of Bohr and Sommerfeld (equation (3.7)) $n = 1, 2, 3$ energy levels for $Z = 1$, in electronvolts. Not to scale; splitting of levels is greatly exaggerated for clarity. Colors are not meaningful, but are consistent with figures 3.1 and 3.2.

$$|d\lambda_{\text{Å}}| \sim 1200\left(\frac{4.5 \times 10^{-5}}{10}\right) \sim 0.0054 \text{ Å}.$$

The 2 → 1 spectral lines will be separated by only a few thousandths of an Ångstrom at around 1216 Å. That such splitting can be detected is a testament to the extreme precision achievable with spectroscopy. If you prefer to use wave numbers in cm^{-1}, you can use equation (1.4):

$$\tilde{\nu} = 8065E_{\text{eV}} \Rightarrow d\tilde{\nu} = 8065(dE_{\text{eV}}) \sim 8065(4.5 \times 10^{-5}) \sim 0.36 \text{ cm}^{-1}.$$

Because spectral lines could be further split by the imposition of external electric and magnetic fields (the Stark and Zeeman effects), Sommerfeld ended up having to add more quantum numbers to his scheme, eventually incorporating four. These modifications lie well beyond the intended scope of this book, however, so let us return to Bohr in the 1910s.

An aspect of an academic career that is often not apparent to students is that a researcher's contributions to some area can proceed in fits and starts as other responsibilities, particularly administrative ones, come and go as an individual moves up through institutional hierarchies. Between 1914 and 1916, Bohr was back at Manchester University while he was waiting for a position to open up at the University of Copenhagen. This materialized in 1916, but facilities and working conditions left much to be desired. Fearing that Copenhagen would be unattractive to bright young researchers, he proposed that the university should develop an institute for fundamental research in theoretical and experimental physics. The wheels of academic bureaucracy and fundraising can turn with glacial slowness, however, and it was not until March, 1921, that the University Institute for Theoretical Physics was inaugurated; in 1965 it would be renamed the Niels Bohr

Institute. Much of Bohr's time between 1917 and 1921 was devoted to planning the facilities and programs of the new organization. In the 1930s, the Institute would begin to receive significant financial support from the Carlsberg Foundation—the same Carlsberg as in the beer.

By the late 1910s, Bohr's scheme of ring-structures for multi-electron atoms was proving inadequate; it was unable to predict the characteristics of even the ground state of helium. Bohr and his collaborators (notably Hendrik Kramers), began moving to the conclusion that helium and other atoms could not be planar structures, but rather might involve something with intersecting three-dimensional orbits, as sketched in figure 3.6.

Around 1920, Bohr returned to the question of the periodic table. This culminated with papers published in *Nature* in March and September, 1921, and then a comprehensive 64-page article in the leading German journal *Zeitschrift für Physik* in March, 1922. In June of that year he gave a series of seven lectures on the theory at the University of Göttingen to audiences which included rising stars in the development of quantum mechanics, including Wolfgang Pauli, Pascual Jordan, and Werner Heisenberg. It was at these lectures that Bohr first met Heisenberg (figure 3.7), with whom he developed a close friendship. Tragically, this friendship would be severed in 1941, a development described in chapter 5. Bohr's Correspondence Principle was a motivating element of Heisenberg's 1925 development of the matrix formulation of quantum mechanics. 1922 ended on an even higher note for Bohr when he was awarded that year's Nobel Prize for physics.

The evolution of Bohr's thinking has been described by Helge Kragh as that there was no single Bohr atom, but rather a succession of models sharing some common features (in Aaserud and Kragh 2015). In his 1921/1922 model, multi-electron atoms involved a system of electrons moving in precessing elliptical orbits of different eccentricities, some of which could penetrate close to the nucleus; this is sketched in figure 3.8, although the structure was actually presumed to exist in three dimensions.

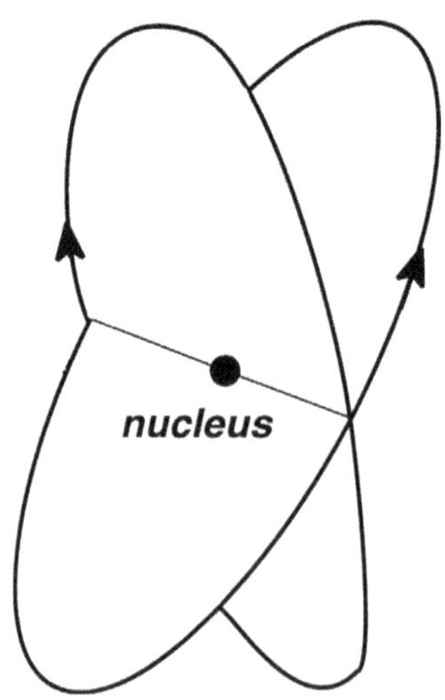

Figure 3.6. Bohr and Kramers intersecting-orbits model of helium.

Figure 3.7. Heisenberg and Bohr, 1934. *Source*: Fermilab, U.S. Department of Energy Public domain; https://commons.wikimedia.org/wiki/File:Heisenbergbohr.jpg.

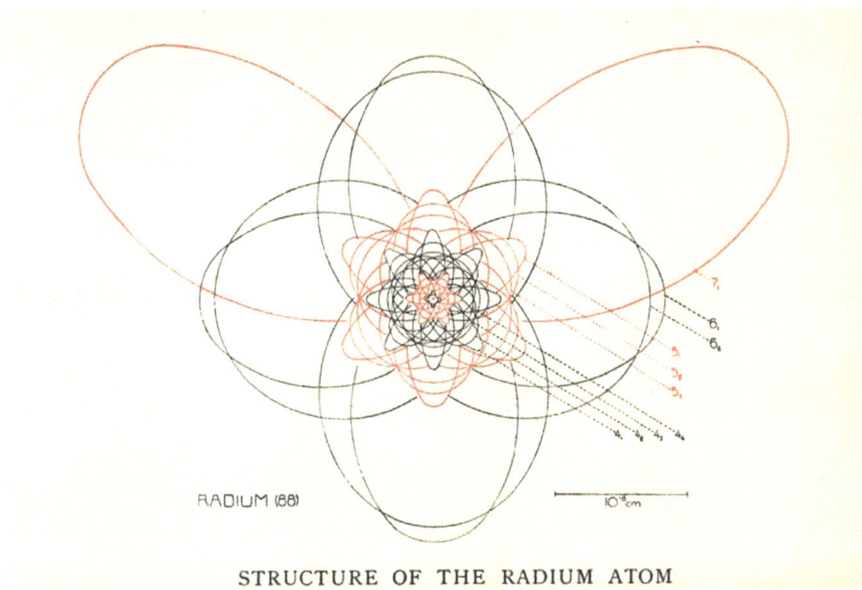

Figure 3.8. Bohr's structure for radium. *Source*: AIP Emilio Segrè Visual Archives.

These orbits involved two quantum numbers, a principal one, n, analogous to that of the original 1913 theory, plus a second, usually designated as k, which dictated the eccentricity as in Sommerfeld's theory.

Surprisingly, these models were not substantiated by extensive calculations, but rather relied on a blend of empirical evidence (particularly spectroscopic) and intuitive considerations, all guided by the Correspondence Principle, which seemed to be a sort of magic wand that operated only in Copenhagen.

Beautiful as these pictures of electron orbits were, they soon ran into conflict with spectroscopic and chemical data. In 1925, the new quantum mechanics began to emerge with Wolfgang Pauli's development of the exclusion principle, and a new guiding philosophy took hold: that theories of atomic structure should be based only on information about observable quantities such as the wavelengths and intensities of spectral lines and chemical properties of elements and molecules, not on attempts to picture structures in the way that characterizes Newtonian physics. By the end of 1926, wave mechanics had replaced Bohr's rings and ellipses with a description of electrons as not being restricted to discrete orbits, but rather as effectively being smeared out into cloudlike distributions centered on the nucleus, with the theory predicting the probability of finding the electron at any point within the cloud.

Some of these probability distributions for hydrogen are shown in figures 3.9(a) and (b). The panels in these images are cuts through the probability distributions in

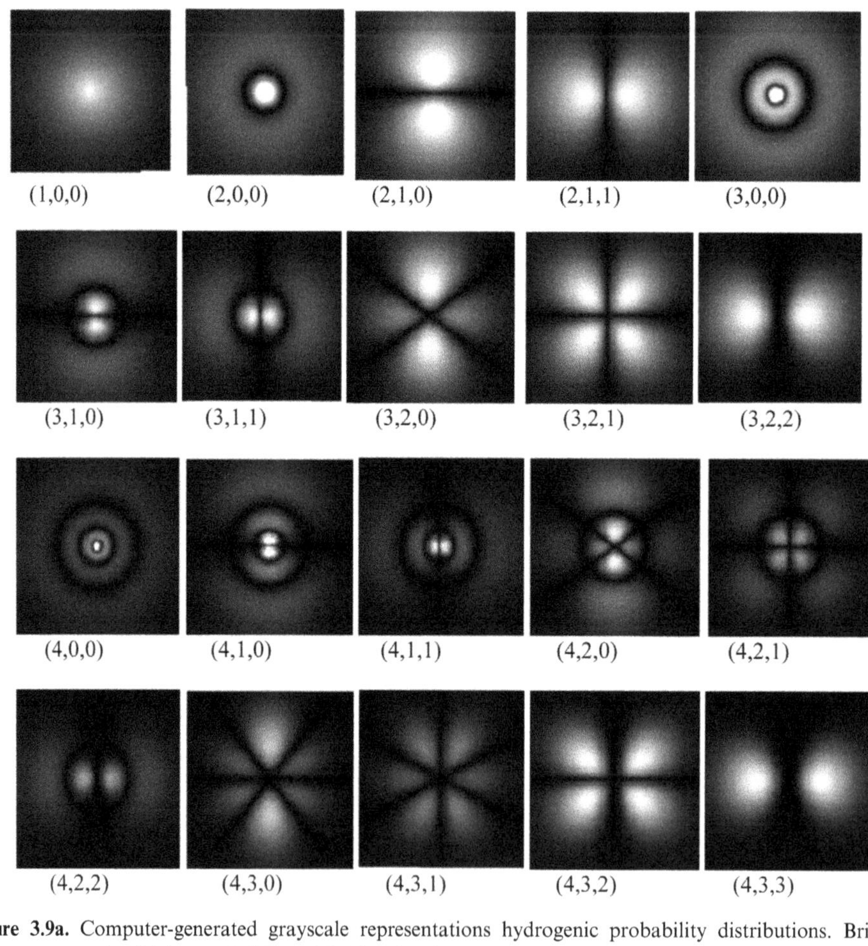

Figure 3.9a. Computer-generated grayscale representations hydrogenic probability distributions. Brighter areas correspond to higher probability of finding the electron at that location. The three numbers below each panel are the (*n, l, m*) quantum numbers of each state.

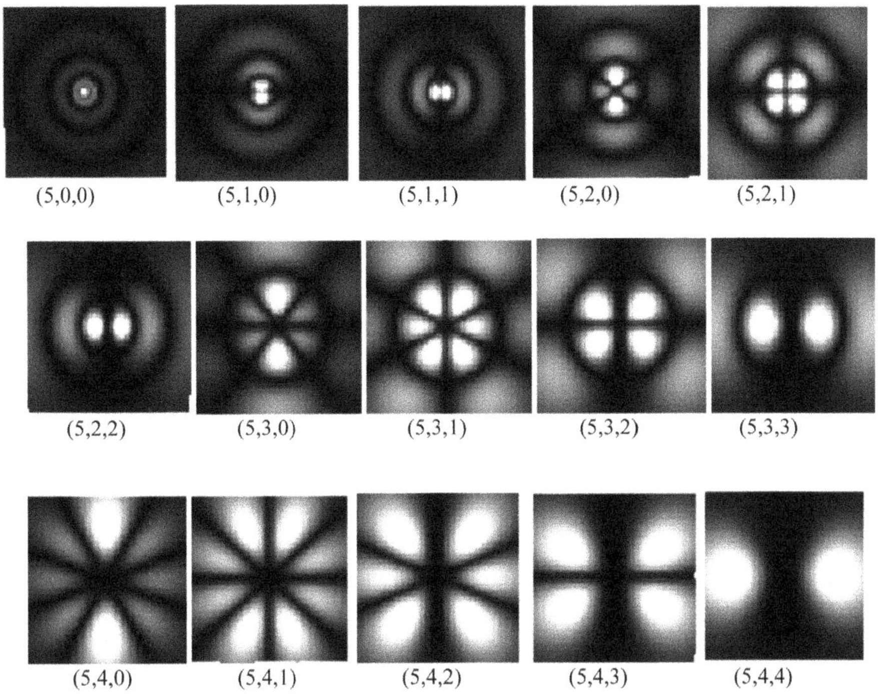

Figure 3.9b. Computer-generated grayscale representations hydrogenic probability distributions. Brighter areas correspond to higher probability of finding the electron at that location. The three numbers below each panel are the (n, l, m) quantum numbers of each state.

the (x, z) plane, with x horizontal across the page and z vertical; the distributions are rotationally symmetric about the z-axis: to envision them in three dimensions, imagine rotating them around an axis going vertically through their centers. In the quantum-mechanical picture, three quantum numbers, traditionally designated (n, ℓ, m), are necessary to fully specify the probability distribution. n is identical to Bohr's orbit number n, and is now called the principal quantum number; the total energy of any (n, ℓ, m) state is identical to that given in equation (2.8). The width of each panel in the figures is $2n^2$ Bohr radii. ℓ and m dictate the shape of the probability distribution. ℓ is known as the angular momentum quantum number; this dictates the total angular momentum of the state, and can run from 0 to $n - 1$; it is somewhat analogous to Sommerfeld's k. When $\ell = 0$, the total angular momentum is zero, and such a distribution is spherically symmetric. m is known as the magnetic quantum number; it can run from $-\ell$ to ℓ. Only pictures for $m = 0$ to ℓ are shown, since the probability distribution for $-m$ is identical to that for positive m. These pictures emerge from the solution of the Schrödinger wave equation, which can be described as the quantum analog of Newton's $F = ma$.

Remarkably, using an experimental technique known as electron momentum spectroscopy, it is possible to measure the distribution of electrons within atoms and molecules, and such measurements prove to be in excellent agreement with theoretical expectations

(Lohmann and Weingold 1981). Any debate as to the meaning of quantum mechanics pales before its success in yielding verifiably correct predictions of atomic behavior.

Bohr was deeply concerned with the philosophical implications and interpretation of the new theory, keeping in close contact with its developers. Heisenberg spent the 1926/27 academic year in Copenhagen, struggling to make sense of the wave and matrix interpretations of the new physics. The situation became even more confusing in 1927, when Heisenberg developed his famous uncertainty principle, which states that the position and momentum of a particle cannot both be known simultaneously with arbitrary precision. To express this briefly and very cursorily, an extended wave has no precise position. To form a highly-peaked distribution that would represent a point can be achieved by adding together waves of different wavelengths, but this corresponds to adding together numerous momenta, so the momentum then has no precise value; recall $p = h/\lambda$ from the discussion of de Broglie matter-waves in section 3.6.

Ideas developed during this period eventually gelled into what has become known as the 'Copenhagen interpretation' of quantum mechanics: that particles can exhibit both wave and particle properties, depending on what method is used to detect them; that Schrödinger wave functions are to be interpreted probabilistically; that the Heisenberg and Schrödinger formulations of quantum mechanics complement each other; and that the predictions of quantum mechanics must merge with those of classical mechanics when quantum numbers are large. Bohr developed the term 'complementarity' to describe the dual wave/particle nature of subatomic particles, concepts which in classical physics were quite divorced from each other. Reminiscences of Heisenberg's time in Copenhagen and his discussions with Bohr can be found in French and Kennedy (1985). Complementarity became a key element of Bohr's personal philosophy; his coat of arms bore the motto *Contraria sunt complementa*: contrary things are complementary. Many of his collaborators recalled that he was fond of saying that truth was complementary to clarity.

Heisenberg's uncertainty principle led to a famous debate between Einstein and Bohr at a conference in October, 1930. A formulation of the uncertainty principle equivalent to its position-momentum statement is an energy-time one: that it is impossible to specify the energy of a system at an arbitrarily precise time. Einstein was uncomfortable with the uncertainty principle, and attempted to devise thought experiments to show that it could be circumvented. (The term 'thought experiment' is used in physics in the sense of an imaginary experiment that could in principle be performed, even if it would be difficult to realize for practical purposes.) Einstein's 1930 thought experiment has no direct relation to the Bohr atom or Bohr's later contributions to physics, but exemplifies the keen insight and ingenuity of both men.

Einstein's experiment is now known as the clock-in-the-box experiment; see figure 3.10. A box filled with electromagnetic radiation (a 'photon gas') and containing a clock is suspended from a spring scale. The clock operates a shutter, which can be opened at a precisely determined time to release one photon from the box. The energy emitted can be determined (in theory, at least!) by weighing the box before and after the emission, and using Einstein's famous equation $E = mc^2$. Both time and the energy of the system then seem knowable to arbitrary accuracy, contrary to the uncertainty principle.

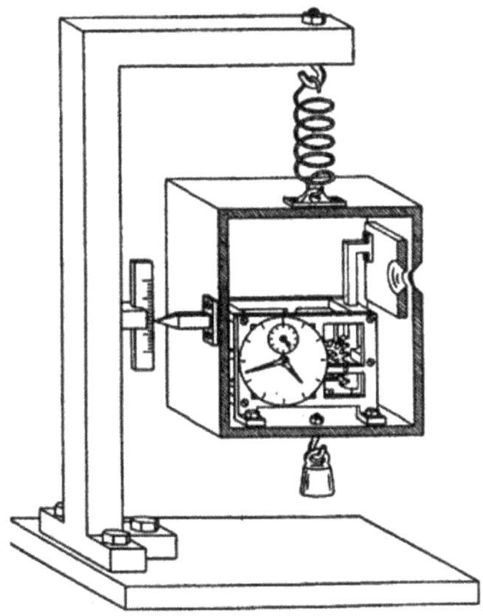

Figure 3.10. Sketch of Einstein–Bohr 'clock-in-the-box' thought experiment. *Source*: Public domain; https://commons.wikimedia.org/wiki/File:Discussion_Bohr_Einstein_Fig_8.gif.

Einstein developed this argument during an evening conversation, greatly disturbing Bohr. He must have had a restless night, for at breakfast the next morning he deployed a fully-formed counterargument, which, ironically, involved Einstein's own 1915 general theory of relativity. This was that weighing the box will involve measuring its position in the Earth's gravitational field. There will be imprecision in the location of the box, and hence its mass, and hence its energy. But even if the spring scale could be read to arbitrary accuracy, there is a further effect: when the box moves after the photon is released, so does the clock, which then ticks in a gravitational field very slightly differently from what it did in its initial position. Einstein's relativity theory held that the passage of time depends on position within a gravitational field, and Bohr was able to show that the corresponding uncertainties in energy and time were as the uncertainty principle demanded. If you are familiar with more advanced quantum mechanics, technical details on this point can be found in Part III of French and Kennedy (1985) and in chapter 19 of Pais (1991). The minute differences in the rate of passage of time over distances of centimeters can now be accurately detected with laboratory-based atomic clocks. Were these effects not programmed into the software in smart phones that report your location to remarkable accuracy, you would not find yourself where they claim you to be.

An implication of the particle/wave duality that goes against every intuition of classical physics is that it is not possible to separate the behavior of atomic objects from the instruments used to measure their properties: does your experiment measure a wave property, or a particle property? You can have one or the other, but not both simultaneously. The universe at the atomic level behaves in ways very counter to our macroscopic intuition. For his part, Einstein remained profoundly disturbed by the implications of quantum mechanics for the rest of his life, maintaining that the theory was somehow incomplete; his debates with Bohr carried on for several years.

Most working physicists are now not particularly concerned with debates regarding the meaning of quantum mechanics, preferring a more functional philosophy of 'Shut up and calculate!', an expression whose origin is itself uncertain (Mermin 2004).

3.9 Bohr's treatment of the H_2 molecule

Diatomic hydrogen is the simplest molecule, and has been well-studied by chemists. In this section, we look at Bohr's attempt to predict the bond length and bond energy of this system.

As described in the preceding section, Bohr envisioned H_2 as two electrons orbiting around an imaginary line joining two protons (figure 3.11). The electrons repel each other, but will be attracted to the protons; similarly, the protons repel each other whilst being attracted to the electrons. The first part of the calculation is to examine the forces acting within the system to see if a stable arrangement can be had. Once this is established, the overall energy of the configuration is analyzed.

Let the orbital radius of each electron be a, and the distance of each proton from the plane of the electron orbits be b. The distance between each electron and each proton is $\sqrt{a^2 + b^2}$, and the angle θ shown in figure 3.11 will have

$$\{\cos\theta, \sin\theta\} = \left\{\frac{a}{\sqrt{a^2 + b^2}}, \frac{b}{\sqrt{a^2 + b^2}}\right\}. \tag{3.8}$$

These expressions will be useful in setting up force components.

Consider first one of the protons, say the topmost one. As sketched in figure 3.12, it will experience a repulsive force F_p from its fellow proton at a distance of $2b$, and an attractive force F_e to each neutron. The sidewise components of the attractive

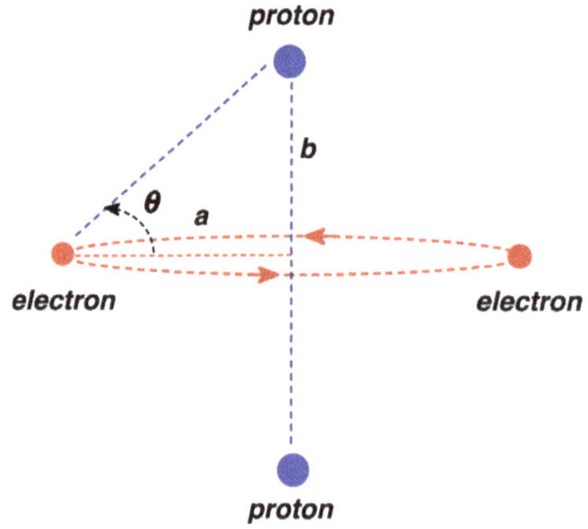

Figure 3.11. Conceptual sketch of Bohr's model of an H_2 molecule. Two electrons orbit the line joining the two protons in a plane perpendicular to the line. Colors are for illustrative purposes only.

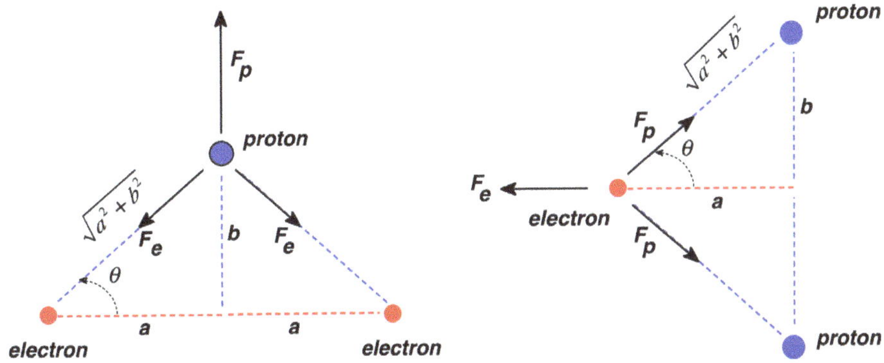

Figure 3.12. Force diagram for proton (left) and electron (right). Not to scale. Colors are for illustrative purposes only.

forces will cancel, but each will contribute a downward component $F_e \sin\theta$ that will counteract the repulsive force.

If the proton is in equilibrium, we must have

$$\frac{e^2}{4\pi\varepsilon_o(2b)^2} - 2\left[\frac{e^2}{4\pi\varepsilon_o(a^2+b^2)}\right]\sin\theta = 0.$$

The first term in this expression is F_p (upwards), and the second is two copies of the downward component of F_e; the term in square brackets is the magnitude of F_e. On canceling various common factors, using equation (3.8), and rearranging, you should be able to show that this condition reduces to

$$(a^2+b^2)^3 = 64b^6.$$

Now suppose that we set $a = \alpha b$. Then $(a^2+b^2)^3$ reduces to $b^6(\alpha^2+1)^3$, and we find $\alpha = \sqrt{3}$. Hence

$$a = \alpha b = \sqrt{3}\,b. \tag{3.9}$$

Our molecule must be wider than it is tall. We will return to this result presently.

We now need to consider the electrons. Look at, say, the one on the left. It will experience a repulsive (leftward, negative) force from the other electron at a distance of $2a$, and attractive forces F_p toward each proton, which will have components $F_p \cos\theta$ to the right in the plane of the orbit. However, the electrons are executing circular motion, so the *net* force on each must be of magnitude $m_e v^2/a$, where v is the orbital speed. Hence we have

$$-\frac{e^2}{4\pi\varepsilon_o(2a)^2} + 2\left[\frac{e^2}{4\pi\varepsilon_o(a^2+b^2)}\right]\cos\theta = \frac{m_e v^2}{a}. \tag{3.10}$$

At this point Bohr, invokes his angular momentum condition for the electrons, and sets $m_e va = h/2\pi$. Solve this expression for v, substitute into equation (3.10), and use equation (3.8) for $\cos\theta$. Pull the physical constants e and ε_o to the right side, and use equation (3.9) to eliminate b. You should find

$$a\left[-\frac{1}{8}+\frac{1}{(1+1/\alpha^2)^{3/2}}\right]=\frac{h^2\varepsilon_o}{2\pi m_e e^2}=\frac{a_o}{2},$$

where a_o is the Bohr radius.

With this, we can solve for the orbital radius a, setting $\alpha=\sqrt{3}$ (you should verify this as well):

$$a=\left(\frac{4}{3\sqrt{3}-1}\right)a_o=0.953\,a_o. \tag{3.11}$$

The electrons' orbital radius is just shy of one Bohr radius. The proton–proton bond length is then $b=(0.953\,a_o)/\sqrt{3}=0.550\,a_o$. In reality, the bond length is ~1.4 Bohr radii, so this model clearly has shortcomings.

Exercise

Show that equation (3.11) follows from equation (3.10).

The physical point alluded to in section 3.8 emerges upon considering the total energy of the system. This comprises four contributions: the potential energy of the electron/electron pair, the potential energy of the proton/proton pair, the potential energy of the four possible electron/proton pairs, and the kinetic energy of the two electrons. Using the usual electrostatic formula $Q_1 Q_2/4\pi\varepsilon_o d$ for the potential energy of two charges separated by distance d, we get

$$E_{\text{total}}=\frac{e^2}{4\pi\varepsilon_o(2a)}+\frac{e^2}{4\pi\varepsilon_o(2b)}-4\left[\frac{e^2}{4\pi\varepsilon_o\sqrt{a^2+b^2}}\right]+2\left[\frac{1}{2}m_e v^2\right]. \tag{3.12}$$

Again eliminate v with the angular momentum condition, invoke equation (3.8), and write equation (3.11) as $a=\gamma a_o$, where $\gamma=0.953$. After some algebra—a good exercise—you should find that equation (3.12) reduces to

$$E_{\text{total}}=E_{\text{H}}\left[\frac{1}{\gamma}+\frac{\alpha}{\gamma}-\frac{8}{\gamma\sqrt{1+1/\alpha^2}}+\frac{2}{\gamma^2}\right], \tag{3.13}$$

where E_{H} is the magnitude of the ground-state energy of a single hydrogen atom, ~ 13.6 eV in equation (2.8). The term in square brackets is dimensionless, so our expression does have the correct units.

Putting in the numbers for α and γ, you should find $E_{\text{total}}\sim -2.20\,E_{\text{H}}$. *Here is the physical point: in this model, a hydrogen molecule will be in a lower energy state, $-2.20\,E_{\text{H}}$, than the total energy of two lone hydrogen atoms, $-2E_{\text{H}}$. This means that* *two hydrogen atoms will prefer to be in a bound state, not dissociated.* The difference in energy between the two configurations, $0.2E_H$, is the bond energy of an H_2 molecule. This corresponds to ~ 2.72 eV. Chemists prefer to quote bond energies in units of kiloJoules (kJ) per mole. For a mole of H_2 molecules, our 2.72 eV corresponds to ~262 400 J, or ~262 kJ. The true bond energy of H_2 is ~436 kJ/mole, so the Bohr value is about 60% of the correct value. But, as Bohr himself remarked, the result is of the correct order of magnitude, and does predict the stability of H_2 over individual hydrogen atoms. For a first stab at molecular theory, this was a remarkable achievement.

Exercise

A photon of energy 2.7 eV or greater can dissociate an H_2 molecule. To what wavelength does this correspond? HINT: equation (1.3). [Ans: ~4590 Å or less]

Exercise

In equation (3.9), consider α to be fixed as $\alpha = \sqrt{3}$; this is necessary to keep the protons stable. Show that by minimizing E_{total} with respect to γ, you get the same result as equation (3.13). This is another example of the energy minimization technique of section 3.7.

References

Aaserud F and Kragh H (ed) 2015 *One Hundred Years of the Bohr Atom: Proceedings from a Conference* (Copenhagen: Royal Danish Academy of Sciences and Letters)

Davisson C and Germer L H 1927 *Phys. Rev.* **30** 707–40

French A P and Kennedy P J (ed) 1985 *Niels Bohr: A Centenary Volume* (Cambridge, MA: Harvard University Press)

Lohmann B and Weingold E 1981 *Phys. Lett.* A **86** 139

Mermin D 2004 *Phys. Today* **57** 10

Michelson A A and Morley E W 1887 *Philos. Mag. Ser. 5* **24** 463–6

Moseley H G J 1913 *Philos. Mag. Ser. 6* **26** 1024–34

Moseley H G J 1914 *Philos. Mag. Ser. 6* **27** 703–13

Pais A 1991 *Niels Bohr's Times, in Physics, Philosophy, and Polity* (Oxford: Oxford University Press)

IOP Publishing

The Bohr Atom
A guide
Bruce Cameron Reed

Chapter 4

Miscellaneous topics

This chapter addresses some miscellaneous topics associated with the Bohr model. These include isotopic effects in spectral lines; the magnetic field created by an electron orbiting a nucleus; how the theory can be used to model the ionization potentials of two-electron atoms; anti-hydrogen; and atomic recoil.

4.1 Isotope effects: deuterium

In 1913, the idea that elements might exist in different isotopic forms was in its infancy. The origin of the concept of isotopy came from chemical analyses of products of radioactive decay chains. This work revealed substances of different atomic masses which could not be separated from each other chemically, that is, they behaved as identical elements. The idea of isotopes was formalized by chemist Frederick Soddy in late 1913; earlier in his career, he had co-discovered natural radioactive transmutation with Ernest Rutherford. Soddy's paper was published in the December 4, 1913 edition of *Nature*, just after publication of the third paper of Bohr's trilogy (Soddy 1913). He argued that the decay-chain evidence suggested that 'the net positive charge of the nucleus (that is, the atomic number) is the number of the place which the element occupies in the periodic table'. Basing his hypothesis on the then-current idea that the electrically-neutral mass in nuclei was a combination of protons and electrons (neutrons had not yet been discovered), Soddy hypothesized that the 'algebraic sum of the positive and negative charges in the nucleus, when the arithmetical sum is different, gives what I call 'isotopes' or 'isotopic elements', because they occupy the same place in the periodic table'. The root 'iso' comes from the Greek word 'isos', meaning 'equal', and the *p* in *tope* serves as a reminder that it is the number of protons which is identical in all isotopes of a given element.

Hydrogen comprises two naturally-occurring isotopes, ^{1}H and ^{2}H. The former is the usual form of hydrogen with a single proton as its nucleus. Nuclei of the latter, 'heavy' hydrogen, now known formally as deuterium, comprise a proton and a neutron bound together. Deuterium is present to an extent of about 0.01% in natural hydrogen, but is very important in the nuclear industry, where 'heavy water', chemical symbol D_2O, is used as a non-neutron-capturing moderator and coolant in nuclear reactors. You can make your coffee with heavy water and your body would digest it as it does normal water, but the coffee would literally be heavier.

Because ordinary hydrogen and deuterium have different-mass nuclei, their reduced masses are slightly different, which means that the wavelengths emitted in various electron transitions will be slightly different. In this section, the difference in wavelength for the Balmer-alpha (3 → 2) transition is calculated.

For ordinary hydrogen (*H*), the expression for the Bohr transition wavelengths is equation (2.14):

$$\lambda_{i \to f} = \left(\frac{8\varepsilon_o^2 h^3 c}{e^4 \mu_{\mathrm{H}}}\right)\left(\frac{1}{n_{\mathrm{final}}^2} - \frac{1}{n_{\mathrm{initial}}^2}\right)^{-1},$$

where the reduced mass is given as usual by $\mu_H = \left[m_e m_p/(m_p + m_e)\right]$. Setting $n_{\mathrm{final}} = 2$ and $n_{\mathrm{initial}} = 3$ gives

$$(\lambda_{3\to 2})_{\mathrm{H}} = \frac{8\,\varepsilon_o^2\, h^3 c}{e^4}\left(\frac{m_e + m_p}{m_e m_p}\right)\left(\frac{1}{4} - \frac{1}{9}\right)^{-1} = \frac{288}{5}\frac{\varepsilon_o^2\, h^3\, c}{e^4}\left(\frac{m_e + m_p}{m_e m_p}\right).$$

For deuterium, replace m_p with $2m_p$ in the reduced mass. Strictly, this is a slight approximation in that we are assuming that protons and neutrons have the same mass, but their masses differ by less than 1%. This gives

$$(\lambda_{3\to 2})_{\mathrm{D}} = \frac{288}{5}\frac{\varepsilon_o^2 h^3 c}{e^4}\left(\frac{m_e + 2m_p}{2m_e m_p}\right).$$

The difference in wavelengths between the hydrogen and deuterium transitions is then

$$(\lambda_{\mathrm{H}} - \lambda_D)_{3\to 2} \sim \frac{288}{5}\frac{\varepsilon_o^2 h^3 c}{e^4}\left[\left(\frac{m_e + m_p}{m_e m_p}\right) - \left(\frac{m_e + 2m_p}{2m_e m_p}\right)\right].$$

$$\sim \frac{288}{5}\frac{\varepsilon_o^2 h^3 c}{e^4 m_e m_p}\left[(m_e + m_p) - (m_e + 2m_p)/2\right] \sim \frac{144}{5}\frac{\varepsilon_o^2 h^3 c}{m_p e^4}.$$

This evaluates to about 1.78 Å, large enough to be detected with a good-quality spectrometer.

Deuterium was discovered in late 1931 by Harold Urey (1893–1981) and his collaborators by evaporating samples of liquefied hydrogen and isolating the last few cubic centimeters of gas so generated (Urey *et al* 1932). This material, which would presumably be slightly concentrated in any heavier isotopes present, was analyzed with high-precision spectroscopy ($\pm$0.02 Å), and weak 'satellite' lines adjacent to the ordinary hydrogen lines were detected. For the 3 → 2 transition, the observed difference in wavelengths was estimated as ~1.82 Å, in close agreement with the above calculation. Urey was awarded the 1934 Nobel Prize for Chemistry for this discovery.

4.2 Magnetic field of the orbiting electron

If you have studied some electromagnetism, you may be familiar with the idea that an electrical current, that is, moving electrical charges, create magnetic fields. In the Bohr model, the orbiting electron creates a current, which must create a magnetic field. In this section, the magnitude of the field at the location of the proton is estimated.

Electrical currents are designated by the symbol i and have units of amperes, equivalent to Coulombs per second. Magnetic fields are usually designated by the symbol B, and have units of Teslas. One Tesla is equivalent to one Newton per ampere-meter (N A^{-1} m^{-1}), which can be expressed in more fundamental units as one kilogram per Coulomb-second, kg C^{-1} s^{-1}.

A standard textbook example is to compute the magnetic field at the center of a circular loop of current i of radius r. The resulting expression is

$$B = \frac{\mu_O i}{2r}.$$

μ_O is a fundamental physical constant known as the permeability of free space; its value is exactly $4\pi \times 10^{-7}$ Tesla-meters per ampere (T m A^{-1}), equivalent to a kilogram-meter per squared Coulomb, (kg m) C^{-2}.

For application to the Bohr model, the electron orbits the nucleus, which is presumed here to be stationary. The goal is to compute the field created by the orbiting electron at the location of the nucleus for an electron in the nth Bohr orbit.

The electrical current will be the magnitude of the electron charge divided by the time for one orbit. The orbital period was calculated in section 1.4:

$$t_{\text{orbit}} = \frac{2\pi r_e}{v_e} = 2\pi r_e \sqrt{\frac{4\pi\varepsilon_o m_e r_e}{Ze^2}} = \frac{4\pi^{3/2}\varepsilon_o^{1/2} m_e^{1/2} r_e^{3/2}}{Z^{1/2} e}.$$

Hence the current is

$$i = \frac{e}{t_{\text{orbit}}} = \frac{Z^{1/2} e^2}{4\pi^{3/2}\varepsilon_o^{1/2} m_e^{1/2} r_e^{3/2}}. \tag{4.1}$$

With $r_e = a_o n^2$, the field evaluates as

$$B = \frac{\mu_O i}{2r_e} = \frac{\mu_O}{2}\left(\frac{Z^{1/2} e^2}{4\pi^{3/2}\varepsilon_o^{1/2} m_e^{1/2} r_e^{5/2}}\right) = \frac{\mu_O}{2}\left(\frac{Z^{1/2} e^2}{4\pi^{3/2}\varepsilon_o^{1/2} m_e^{1/2}}\right)\left(\frac{\pi Z\, m_e e^2}{\varepsilon_o h^2 n^2}\right)^{5/2},$$

which reduces to

$$B = \frac{\mu_O \pi m_e^2 e^7 Z^3}{8\varepsilon_o^3 h^5 n^5}.$$

You should check that the units do reduce to Teslas. For $Z = 1$ and $n = 1$, this gives a field of ~12.5 Teslas. This is a remarkably strong (although very localized!) magnetic field: the natural magnetic field of the earth at its surface depends on your location, but is only of magnitude ~25–65 μT! A Tesla is far too large a unit of field for terrestrial measurements, so one usually encounters units of Gauss; 1 Tesla = 10 000 Gauss.

A quantity which you may encounter is the *Bohr magneton*, a measure of the magnetic dipole moment of the orbiting electron. This quantity is symbolized as μ_B, and is defined to be the current created by electron when in the $n = 1$ orbit times the area of the orbit, πa_o^2. From equation (4.1) and the Bohr radius,

$$\mu_B = i\pi a_o^2 = \left(\frac{e^2}{4\pi^{3/2}\varepsilon_o^{1/2}m_e^{1/2}a_o^{3/2}} \right) \pi a_o^2 = \left(\frac{e^2 a_o^{1/2}}{4\pi^{1/2}\varepsilon_o^{1/2}m_e^{1/2}} \right) = \frac{eh}{4\pi m_e}.$$

This evaluates to 9.274×10^{-24} A m^2. You should be able to show that this is equivalent to Joules per Tesla: the magneton is a measure of the energy an magnetic dipole will acquire when in the presence of an external magnetic field.

Atomic-scale magnetic fields are important in modern technology. In magnetic resonance imaging (MRI) scanners, powerful magnets force the magnetic moments of protons into parallel alignment. A radio-frequency pulse is then passed through the patient, which forces the protons out of alignment with the field. After the pulse is turned off, the protons realign with the field, releasing energy, which can be detected. The time required for this to happen, typically fractions of a second, depends on the type of tissue involved, and it is possible to discriminate healthy tissue from, say, cancerous material. Remarkably, the entire procedure is non-invasive, and can produce very high-resolution images.

4.3 Ionization potentials of two-electron atoms

While the Bohr model can be applied to atoms from which all but one electron has been ionized, it is awkward to apply to multi-electron atoms because of the presence of inter-electronic forces. However, it is possible to apply it in a straightforward way to an idealized model of two-electron atoms for the purpose of estimating their ionization potentials (IPs), that is, the total amount of energy necessary to remove both electrons. As will be seen in what follows, the results are in surprisingly good agreement with measured values. While this calculation gives no predictions of spectral properties, it does serve to reinforce Bohr's approach and extend it beyond the usual hydrogenic-atom case. The model presented in this section is adopted from Reed (2005).

The model is shown in figure 4.1. Two electrons of charge $-e$ are assumed to be in circular orbits of radius r and speed v_e on opposite sides of a nucleus of charge $+Ze$. Assuming that the two electrons remain opposite each other, the nucleus will remain at the center of mass of the system, so there is no motion of the nucleus to worry about.

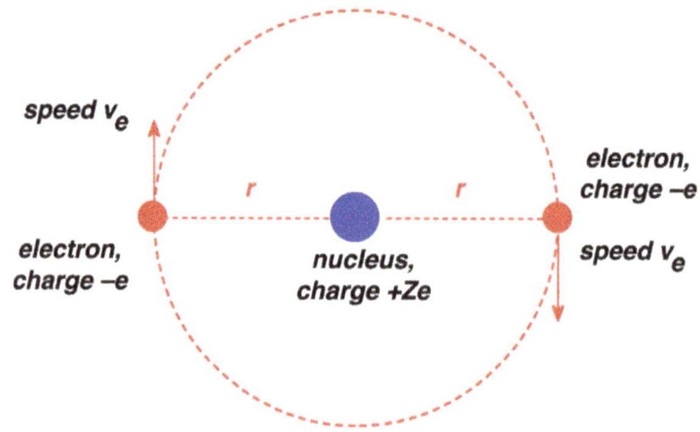

Figure 4.1. Schematic model of a two-electron atom. Colors are for illustrative purposes only.

Consider one of the electrons. The net Coulomb force it experiences is the sum of an attractive force toward the nucleus and a repulsive one from the other electron. The net magnitude of this combination toward the nucleus is then

$$F = \frac{Ze^2}{4\pi\varepsilon_o r^2} - \frac{e^2}{4\pi\varepsilon_o (2r)^2} = \frac{e^2}{4\pi\varepsilon_o r^2}(Z - 1/4).$$

For a circular orbit, the net force must be centripetal and of magnitude $m_e v_e^2/r$, hence

$$v_e^2 = \frac{e^2}{4\pi\varepsilon_o m_e r}(Z - 1/4). \tag{4.2}$$

The potential energy of the system comprises two electron/nucleus contributions and one electron/electron contribution:

$$\text{PE} = \frac{-2Ze^2}{4\pi\varepsilon_o r} + \frac{e^2}{4\pi\varepsilon_o (2r)} = \frac{e^2}{8\pi\varepsilon_o r}(-4Z + 1). \tag{4.3}$$

Now invoke Bohr's angular momentum quantization rule: that the angular momentum $m_e v_e r$ of each orbiting electron is equal to $h/2\pi$; that is, $v_e = h/2\pi m_e r$; we assume that we are dealing with the $n = 1$ state. Squaring this and substituting into equation (4.2) gives the orbital radius as

$$r = \frac{\varepsilon_o h^2}{(Z - 1/4)\,\pi m_e e^2} = \frac{a_o}{(Z - 1/4)}. \tag{4.4}$$

The ground-state radius of our two-electron atom is a little greater than the Bohr value for any Z, a consequence of the repulsive force between the electrons in addition to the usual electron/nucleus attractive force. With equations (4.2)–(4.4), we can compute the total energy of the system:

$$E = \text{KE} + \text{PE} = 2\,(1/2 m_e v_e^2) + \text{PE} = -\left(\frac{m_e e^4}{4\varepsilon_o^2 h^2}\right)(Z - 1/4)^2. \tag{4.5}$$

Table 4.1. Predicted and observed ionization potentials (in eV)[a]. See equation (4.5).

Element	Z	IP (Predicted)	IP (Experimental)	% error
He I + II	2	−83.3	−79.0	5.4
Li II + III	3	−205.8	−198.1	3.9
Be III + IV	4	−382.7	−371.6	3.0
B IV + V	5	−614.0	−599.6	2.4
C V + VI	6	−899.7	−882.1	2.0
Na X + XI	11	−3145	−3114	1.0
Ca XIX + XX	20	−10 599	−10 614	0.14

[a]Experimental values from Lang K R 1980 *Astrophysical Formulae* (Berlin: Springer-Verlag) p 246.

E should correspond to the IP for the last two electrons of any nucleus with $Z \geq 2$; $(m_e e^4/4\,\varepsilon_0^2\,h^2)$ is twice the usual hydrogenic IP of 13.6 eV.

Predicted and actual IPs for a few light elements stripped of all but two electrons are shown in table 4.1; the first column in the table gives the ionization states whose individual IPs were added to compute the experimental values quoted in the second-last column. For example, sodium (Na) normally has 11 electrons. If all but two have been removed, we will be dealing with Na X according as the Roman-numeral notation described in section 2.4. The total IP will then be the sum of those for Na X and Na XI, as indicated in the table.

The results are in surprisingly good accord with experimentally-measured IPs. The trend to decreasing errors with increasing Z is somewhat surprising since the orbital speed of the electrons—hence relativistic effects—will grow as $(Z - 1/4)$ (equations (4.2) and (4.4)); for calcium, $v/c \sim 0.15$. Conversely, the presumably less-uniform charge distributions within lower-Z nuclei would render them less well described by purely Coulomb potentials.

4.4 Anti-hydrogen

A fundamental tenet of modern physics is that all particles possess anti-particles. Particle/antiparticle pairs have the same properties except for having opposite electrical charges. We can then imagine *anti*-hydrogen: an anti-electron (known as a positron) in orbit around an anti-proton. Modern particle accelerators generate copious quantities of anti-particles, but when matter and anti-matter come into contact, they annihilate each other into pure energy. Fortunately, it is possible to trap anti-particles in magnetic fields for hours, long enough to perform experiments with them. Anti-hydrogen was first produced at the CERN research laboratory in Europe in 1996, and in 2016, scientists there managed to measure the n = 1–2 transition, finding it to be identical to that of 'ordinary' hydrogen to a precision of about one part in ten billion (Ahmadi *et al* 2017). Modern particle physics theories hold that the spectra of the two should be identical; any deviation would have pointed to fundamental problems with the theory.

4.5 Atomic recoil

The calculation done here is not a Bohr model issue *per se*, but does relate to electron transitions.

As described in section 3.6, photons carry linear momentum according as

$$p = \frac{h}{\lambda}.$$

When a photon is emitted in a transition, it must carry off some momentum. But total momentum is conserved for an isolated system, which means that the parent atom (mass m) must recoil with some speed v in the opposite direction of the photon such that $mv = h/\lambda$, which means that the atom will carry off some kinetic energy. Consequently, not all of the energy liberated in the transition can go with the photon, a fact which was completely ignored in the derivation of the Bohr model. This section looks at the ratio of the atomic recoil to photon energies. A classical analysis will suffice for the recoil.

The recoil speed must be

$$v = \frac{h}{m\lambda}.$$

The kinetic energy carried off in the recoil is then

$$K_{\text{recoil}} = \frac{1}{2}mv^2 = \frac{h^2}{2m\lambda^2}.$$

The ratio of the recoil kinetic energy to the energy of the photon is then

$$\frac{K_{\text{recoil}}}{E_{\text{photon}}} = \frac{h^2}{2m\lambda^2}\left(\frac{\lambda}{hc}\right) = \frac{h}{2m\lambda c}.$$

For a recoiling hydrogen atom ($m = 1.67 \times 10^{-27}$ kg) and a photon of $\lambda = 6500$ Å, the recoil speed is a very leisurely 0.6 meters per second, and $K_{\text{recoil}}/E_{\text{photon}} \sim 1.0 \times 10^{-9}$. It is certainly justifiable to treat the recoil as non-relativistic, and to neglect the kinetic energy carried off by the atom in comparison to the energy of the photon. This is a tiny effect, but it has been measured (Campbell *et al* 2005).

References

Ahmadi M *et al* 2017 Observation of the 1s–2s transition in trapped antihydrogen *Nature* **541** 506–10

Campbell G K, Leanhardt A E, Mun J, Boyd M, Streed E W, Ketterle W and Pritchard D E 2005 *Phys. Rev. Lett.* **94** 170403

Reed B C 2005 *Phys. Educ.* **40** 117–8

Soddy F 1913 *Nature* **92** 399–400

Urey H, Brickwedde F and Murphy G 1932 *Phys. Rev.* **39** 164–5

Chapter 5

The 1930s, World War II, fission, the bomb, and politics

This closing chapter reviews Bohr's life and work from the late 1920s until his death in 1962, focusing in particular on his activities during and after World War II in conjunction with the development of nuclear weapons and arms control efforts. For more detail, readers are urged consult any of the excellent full-scale and article-length biographies of Bohr, for example, those of Aaserud and Heilbron (2013), Aaserud and Kragh (2015), Ford (2018), French and Kennedy (1985), Heilbron and Kuhn (1969), Kragh (2011, 2012), Pais (1991), and Schwarz (2013).

By the late 1920s, the structure of atoms and their corresponding chemistry was becoming understood on the basis of wave mechanics. Attention began to turn to nuclei and the emerging field of nuclear physics. Here again Bohr made seminal contributions, particularly in what came to be known as the compound nucleus model of nuclear reactions.

This model was developed in the latter half of 1935, and was published in early 1936 (Bohr 1936). This work considered nuclear reactions, particularly ones wherein bombarding neutrons are captured by nuclei to become compound entities. Some mass-energy is typically released in these reactions; the nucleus becomes agitated, and then sheds this energy in one or more of a number of possible ways, such as the emission of a proton, neutron, alpha-particle or gamma-ray. Bohr pictured this as akin to a game of billiards, with a neutron propelled into a nucleus from the outside world (figure 5.1). The concept of a compound nucleus would play a key role in the later development of a theory of fission.

This golden age of physics began to unravel with the economic crash of 1929 and the rise of Fascism in Europe. Soon after Adolf Hitler became Chancellor of Germany in early 1933, anti-Semitic laws forced over 1000 scholars in that country from their teaching positions. Even while continuing to pursue physics, Bohr, deeply pacifist and humanistic, worked to secure positions for many of these exiles in Scandinavia, Britain, and particularly America, where several of them made

doi:10.1088/978-0-7503-3611-6ch5

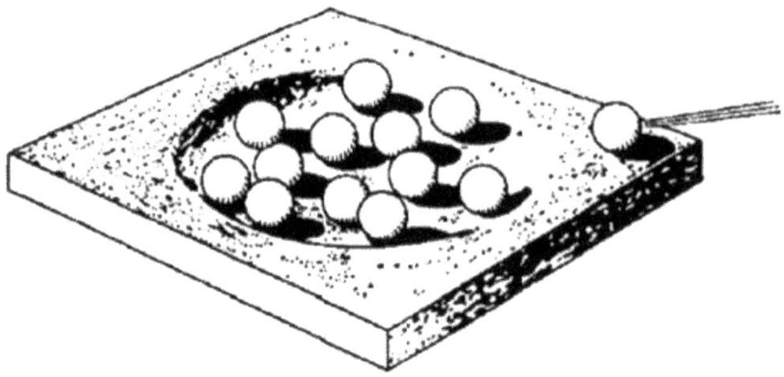

Figure 5.1. Bohr's compound nucleus model. *Source*: Hullernuc commons.wikimedia.org; https://commons.wikimedia.org/wiki/File:Compnuc.png.

significant contributions to the atomic bomb project during World War II. Soon after Hitler came to power, Bohr made a trip to Germany to let friends and colleagues know that they would be welcome in Copenhagen, and he and his brother Harald helped to organize the Danish Committee for the Support of Refugee Intellectuals. He also helped Enrico Fermi and his family escape to the United States following their travel to Stockholm in December, 1938, when Fermi was awarded that year's Nobel Prize for Physics.

The world of nuclear physics was shaken with the announcement, in early 1939, of the discovery of uranium fission, the process wherein nuclei of that element can be caused to break into two upon being bombarded by a neutron. No such reaction had ever been detected before in over 40 years of nuclear research. The stunning thing about this process is that the mass of the fragments that emerge from the fission total to slightly less than that of the original nucleus plus the incoming neutron by about one-fifth of the mass of the proton. While this may seem tiny, it corresponds to about 200 *million* electronvolts of energy via $E = mc^2$, millions of times more than is typical of chemical reactions; the energy appears as (mostly) the kinetic energy of the fragments. The explosion of one kilogram of uranium 235 (see below) in this way releases an amount of energy equivalent to detonating some 17 000 metric tons of ordinary chemical explosive: 17 'kilotons'. Within a few weeks of the discovery, it was further realized that a by-product of each fission is the prompt release of two or three neutrons, which can potentially go on to strike other uranium nuclei and so initiate an immensely powerful chain-reaction explosion.

Bohr learned of the discovery shortly before leaving Copenhagen for an extended visit to the Institute for Advanced Study (by then Einstein's home base) in Princeton, New Jersey, and instantly understood the implications. During his journey, Bohr began to work out a theory of fission, work he continued in collaboration with John Wheeler of Princeton University; they had known each other since 1934, when Wheeler had been a postdoctoral student in Copenhagen. The result of this collaboration was a monumental 25-page paper published in *The Physical Review* on September 1, 1939, the opening day of World War II (Bohr and Wheeler 1939). Bohr and Wheeler's analysis explained why fission could only occur in the heaviest elements, and also that it was the 235 isotope of uranium that was responsible for fission, not the 238 isotope. The numbers here refer to the total number of (neutrons + protons) in the nuclei concerned. All uranium atoms have 92 protons, so the 235 isotope has 143 neutrons, while the 238 isotope has 146. To make an explosive chain reaction—a nuclear bomb—would require isolating kilograms of U-235 atoms. The catch was that U-235 comprises only about 0.7% of natural uranium, with U-238 comprising 99.3%. Much of the $2 billion cost of the wartime Manhattan Project was devoted to precisely this problem. The result was the uranium-fueled *Little Boy* bomb dropped at Hiroshima on August 6, 1945. An excellent survey of the Manhattan Project can be found in Reed (2020). To some extent, Bohr's remark about turning the United States into one huge factory cited at the start of this book is exactly what was done. After the war, Wheeler would go on to a distinguished career, particularly in the area of cosmology; in 1967 he coined the term 'black hole'.

Bohr returned to Copenhagen in the spring of 1939, where he continued to work on various aspects of fission and nuclear energy. After the German occupation of Denmark in April, 1940, he had to hastily destroy all records of refugee scientists whom he had aided. An especially fateful episode occurred in September, 1941, when Werner Heisenberg visited Copenhagen to speak at the German Scientific Institute, a German propaganda outlet. Sometime during his visit, Heisenberg had a private conversation with Bohr. Unfortunately, there is no contemporary record of their exchange; what we know of what happened can only be reconstructed from letters and comments that each made after the war.

Some context for this meeting is relevant. Despite his high position in the German nuclear program overseeing experiments aimed at constructing a reactor, Heisenberg was politically vulnerable. In 1936, he had been accused in an SS publication of being a 'White Jew' for his support of 'Jewish physics', a code phrase for Einstein's theory of relativity and quantum physics. Fortunately, Heisenberg's mother and SS leader Heinrich Himmler's mother moved in the same social circles, and the former relayed a personal letter from Heisenberg to Himmler asking whether he approved of such attacks. After a year-long SS investigation, Himmler indicated that the attacks would cease. In such circumstances, Heisenberg could hardly have refused the government's 'invitation' to visit Copenhagen; this was one of several such travels he made during the war years. While Heisenberg apparently personally abhorred the Nazis and hoped that Hitler would be replaced as soon as the war was won, he shocked colleagues during a visit to Holland in 1943 by stating that he felt that since democracy could not possibly rule Europe, the continent faced a choice between Germany and the Soviets, with the former clearly to be preferred.

Heisenberg's version of the meeting, as related in a 1948 document, was that he wanted to ask Bohr 'Does one as a physicist have the moral right to work on the practical exploitation of atomic energy?' Historian of science David Cassidy, who has studied this affair closely, speculates that Heisenberg may have hoped to have Bohr use his influence with Allied scientists to prevent them from working on a bomb which could be used against Germany; similarly, the distinguished nuclear physicist Hans Bethe has stated that he believed that Heisenberg might have been trying to tell Bohr that the Germans were working on reactors, not bombs, so that Bohr could be a 'messenger of conscience' to persuade Allied scientists to also refrain from working on bombs (Cassidy 2009, 2017, Bethe 2000). In contrast, in a draft letter to Heisenberg written after the war but never sent, Bohr claimed that he distinctly recalled Heisenberg asserting that if the war lasted long enough, it would be decided with atomic weapons.

Exactly what happed next is unclear, but Bohr apparently became very agitated and asked Heisenberg if an atomic weapon was truly possible. In response, Heisenberg may have given or shown Bohr a diagram of a reactor, which would have been a treasonable act which could have put both men's lives in jeopardy. Whatever transpired, Bohr evidently interpreted the situation as an indication that the Germans were working on atomic weapons, and abruptly terminated the conversation.

After the war, Bohr drafted several letters to Heisenberg giving his side of the story, but never sent them. The Bohr family released the letters in 2002; they are available at the Niels Bohr Archive, accompanied by translations into English, but they do not clarify the situation so far as the diagram is concerned (Niels Bohr Archive 2002). Sadly, the two men never fully reconciled, although some letters were exchanged after the war and they did occasionally meet at conferences. If Heisenberg's intent was benign and he was misinterpreted by Bohr, it is hard to imagine the evolution of events on the Allied side being different from what they were even if Bohr had tried to bring a message that the Germans were not working toward a bomb. He eventually took the drawing to Los Alamos (or redrew it from memory), where it was considered at a meeting held on December 31, 1943. Confusion prevailed as to whether it represented a reactor or a bomb; Bohr was apparently still unclear on the distinction between how the two types of devices operate. But the group who examined it wondered if their German counterparts had gone crazy and were planning to drop a reactor on London! The story of the drawing and its many internal inconsistencies is related in more detail in Sweet (2002) and Bernstein (2003). Bohr's son Aage has claimed that no drawing was exchanged in 1941, and Bernstein suggests that the Los Alamos drawing might have originated in a 1943 visit to Bohr by another German scientist, Hans Jensen. His argument is compelling: the Germans had not begun work on the reactor design sketched in the purported 1941 drawing, but did so in 1942. Heisenberg's ill-fated visit inspired a popular play, *Copenhagen*, the script of which has been published as a book (Frayn 2000).

During the Nazi occupation, Bohr initially resisted offers to be spirited to Britain or the United States. However, Nazi suppression grew as resistance sabotage became more daring. Finally the Germans ordered arrests of Danish Jews, including Bohr, who was tipped off and escaped in a fishing boat with his wife to Sweden in September, 1943; their children followed a few weeks later. Bohr was then flown to Scotland in a dangerous operation. Outfitted with a flight suit and parachute, he was transported in the cramped bomb bay of a Mosquito bomber, a flight which took three hours. To avoid German flak, which could reach to 20 000 feet, the bomber flew high, and Bohr was supposed to wear an oxygen mask. However, his head was too large to fit into the radio-equipped helmet he had been given, and he passed out when he did not hear the pilot's order to don the mask. When the pilot could not raise him and got beyond the range of German guns, he descended, and Bohr regained consciousness.

Bohr was soon taken to London, where he was briefed on the American-British atomic bomb project. Despite efforts to keep his whereabouts secret, the *New York Times* reported on October 9 that Bohr ' … reached London from Sweden … bearing … plans for a new invention involving atomic explosions'. In November, he and Aage departed for the United States, where they would spend much of the war. Soon after Christmas they visited Los Alamos, by which time Bohr was fully aware of just how far his 'one huge factory' had progressed. Formally, Bohr was a member of a contingent of British scientists attached to the Manhattan Project, assigned the code name 'Nicholas Baker'. His pacifist sympathies and belief that Russia should be informed of the bomb guaranteed his surveillance by security agents, but there is absolutely no evidence that he did anything inappropriate. One memorandum from the project's head of security to a top administrator included a plea that Bohr and his son be advised to be more careful when absent-mindedly wandering through Washington traffic. In any case, the Russians did not need Bohr's help: they had several well-placed sympathizers at Los Alamos who were passing information, and were well-informed of what was happening.

Bohr made several visits to Los Alamos during the war, and, while he was involved with some technical work, he became regarded as an elder statesman of science and the voice of conscience of the work. That he worked on the bomb may seem incongruous given his pacifist beliefs, but, like many others who did so despite their moral concerns, the idea of Hitler armed with nuclear weapons to which there would be no comparable response was too terrifying to contemplate.

When it became clear that the project would likely succeed, Bohr began to become concerned with the impact of nuclear weapons on the world political situation, fearing that a destabilizing arms race could emerge after the war. As a result, he began to formulate his thoughts into a proposal for an open world where secrecy would be removed and international exchange of information would be encouraged to secure a lasting peace, which would be supported by some form of international control of nuclear energy.

Bohr felt that in order to pave the way for future cooperation and control, Russia would have to be let in on the existence of the bomb project, and presented this idea in personal meetings with British Prime Minister Winston Churchill and President Franklin Roosevelt. As might be expected, his idealism ran up against the hard realities of politics and strategy. Churchill had already indicated to British officials overseeing their end of the bomb project that he would consider no possibility of collaboration with Russia on any account, but he did agree to meet with Bohr on May 16, 1944. The meeting was a disaster. Churchill was in a bad mood, and Bohr had a frustratingly convoluted way of speaking, often mumbling and hesitantly interrupting himself with new ideas as he went along, a manifestation of his belief that one should never speak more clearly than one thinks. For Churchill, consumed with preparations for the D-Day invasion of France just three weeks in the future, Bohr's professorial strategizing must have seemed naïve at best and a dangerous security threat at worst. Churchill did maintain an interest in scientific developments, but did not yet seem to appreciate the profound strategic changes that nuclear weapons would portend, telling Bohr that 'After all this new bomb is just going to be bigger than our present bombs. It involves no difference in the principles of war. And as for any post-war problems there are none that cannot be amicably settled between me and my friend, President Roosevelt' (see Rhodes 1986, p. 530). Bohr related to an acquaintance after the meeting that Churchill had scolded him as if he were a schoolboy. Churchill's fierce belief that the project be kept secret was probably motivated by his understanding that Britain would emerge from the war as a power much reduced from what it historically had been.

One of Bohr's contacts in Washington was Felix Frankfurter, a Supreme Court justice and confidant of President Roosevelt. Frankfurter arranged a private meeting between Roosevelt and Bohr, which occurred on August 26. For weeks in advance, Bohr worked his thoughts into a memorandum, which Frankfurter passed to Roosevelt. The two men met privately and cordially for an hour and a half, with Roosevelt apparently agreeing that an approach to the Soviet Union must be tried. Bohr came away from the meeting greatly encouraged, but his mission had been fundamentally hopeless from the outset. Roosevelt was a master politician, very adept at telling listeners what they wanted to hear. He and Churchill had no intention of letting Russia benefit from a project which had cost $2 billion and over two years of difficult work. The previous August, they had signed a secret agreement stipulating that neither side would employ nuclear weapons against the other, nor pass information to other countries without the consent of the other, positions that were reiterated soon after the Bohr–Roosevelt meeting. Roosevelt and Churchill were surely thinking of the bomb not only in relation to the immediate wartime situation, but also the power balance of the post-war world and the formidable strength of the Red Army. In the end, given Russian knowledge of the Manhattan Project and the fact that the Russians had already initiated their own atomic program, any sort of approach would likely have made no difference at all. Bohr had hoped to secure another meeting with Roosevelt, but the President's death on April 12, 1945, foreclosed any such possibility.

When the existence of the project was revealed with the bombings of Hiroshima and Nagasaki, Bohr was ready to take his ideas public. Just two days after the bombing of Nagasaki, his essay 'Energy From the Atom: An Opportunity and a Challenge' graced the editorial page of the *London Times*. In what we would now call an op-ed piece, Bohr urged the peoples and countries of the world to unite and cooperate to reap the benefits of nuclear technology, a situation which would require the world's powers to be willing to open up and relinquish some sovereignty to build mutual trust. Proposals to this effect were put forth to the United Nations in 1946, but foundered against the growing mutual suspicion and emerging cold war between America and Russia. As to the use of the bombs to help end the war, Bohr never made much in the way of any public statements.

Bohr returned to Denmark on August 25, 1945, and immediately resumed work at his Institute. At an age when most people begin to wind down their commitments, he essentially began a second career, resuming research and becoming a beloved public figure, giving numerous in-person and radio-broadcast addresses and lectures.

His Institute was expanded and new equipment was installed. His last research paper, on the passage of charged particles through matter, was published in 1954.

Bohr's devotion to the concept of an open world remained with him for the rest of his life. One June 9, 1950, he released a document titled 'Open Letter to the United Nations', a six-page manifesto wherein he addressed himself to what he called the 'adjustment of international relations required by modern development of science and technology'. He began by reviewing some of his wartime efforts, including arguments made in his 1944 memorandum to President Roosevelt, and then moved on to reiterate his belief in the need for free access to information, suggesting that those countries which had pioneered nuclear developments would be in a special position to take some initiative by a direct proposal of mutual openness. Eloquent as his prose was, Bohr offered no practical suggestions on how such a situation might be realized, leaving it for the UN to debate the details, which never happened. However, his cause was essentially hopeless, as the cold war was becoming even more fraught. Russia had detonated its first atomic bomb in August, 1949, and the Korean war broke on June 25, 1950, just two weeks after release of his document. By 1952, immensely powerful hydrogen bombs would be a reality. He tried again with a second letter in 1956, but again events intervened: the Soviet Union brutally repressed a popular uprising in Hungary, and a crisis erupted over the Suez Canal when Israel invaded Egypt in response to that country's nationalization of the Canal. The Israelis were supported by British and French forces, but the three countries were warned by the United States that they would face economic sanctions if they persisted in their attack. The British and French withdrew, and Israel ceded control of the Canal to Egypt.

Bohr was far more successful in science diplomacy, where he was deeply involved with three major initiatives. The first of these was CERN, the European Organization for Nuclear Research (*Organisation européenne pour la recherche nucléaire*; established 1952), which now boasts the largest particle accelerator in the world and was the birthplace of the World Wide Web; Bohr inaugurated the organization's proton accelerator in February, 1960. Closer to home, he was involved with the establishment of NORDITA, the Nordic Institute for Theoretical Physics, which is headquartered in Stockholm (*Nordisk Institut for Theoretisk Atomfysik*; 1957), and the Danish Atomic Energy Commission, which established a laboratory at Risø, not far from Copenhagen. The latter facility's first reactor went critical in August, 1957, and Bohr spoke at the official inauguration in June, 1958. He maintained what must have been an exhausting schedule until just months before his death, with travels often associated with being granted honorary degrees. An incomplete list includes visits to England, Belgium, Scotland, Iceland, Israel, America, Russia, Austria, Yugoslavia, India, and Germany. During a visit to Germany in late June, 1962, he suffered a minor cerebral hemorrhage, but recovered sufficiently to work on a transcript of a lecture he had given there.

Bohr died suddenly of heart failure on the afternoon of November 18, 1962, just a few weeks after his 77th birthday. Tributes poured into Copenhagen from former colleagues and many public figures, including President John F Kennedy, who wrote to Margrethe Bohr that 'American scientists, indeed all American citizens … have respected and venerated him for more than two generations … We are forever indebted to him for the scientific inspiration he brought along on his many visits to the United States, and especially for his great contributions to the atomic center at Los Alamos. Please accept my condolences and deep sympathy'. Newspapers around the world editorialized his scientific genius and deep humanity.

At the time of his death, Bohr was preparing for a conference to be held in 1963 to celebrate the 50th anniversary of his first papers on atomic structure. The meeting was held in July of that year as a memorial, with many of his former collaborators in attendance, including Werner Heisenberg. In 1980, another gathering was held to mark Margrethe Bohr's 90th birthday, with a dinner attended by more than 50 guests. She died on December 21, 1984; her remains were placed next to his.

Bohr's career was remarkable, involving fundamental contributions to physics spanning phenomena from the atomic to the nuclear, complemented with deep consideration of the meaning of physical concepts, the relationship between science and society, and concern for the condition of humanity. It is no exaggeration to say that Bohr did for our understanding of atomic structure what Copernicus, Kepler, and Newton did for our understanding of celestial mechanics. It is rare for any generation to see such a mind. Bohr's name will appear in physics texts for as long as there is physics, and, if humanity becomes sufficiently enlightened, an even better legacy would be that future generations consider him an ahead-of-his-time prophet of the more open world of which he dreamed.

References

Aaserud F and Heilbron J L 2013 *Love, Literature, and the Quantum Atom* (Oxford: Oxford University Press)

Aaserud F and Kragh H (ed) 2015 *One Hundred Years of the Bohr Atom: Proc. from a Conf.* (Copenhagen: Royal Danish Academy of Sciences and Letters)

Bernstein J 2003 *Phys. Perspect.* **5** 243–61

Bethe H 2000 *Phys. Today* **53** 34–6

Bohr N 1936 *Nature* **137** 344–8

Bohr N and Wheeler J A 1939 *Phys. Rev.* **56** 426–50

Cassidy D C 2009 *Beyond Uncertainty: Heisenberg, Quantum Physics, and the Bomb* (New York: Bellevue Literary Press)

Cassidy D C 2017 *Farm Hall and the German Atomic Project of World War II: A Dramatic History* (Berlin: Springer)

Ford K 2018 *Phys. Teach.* **56** 500–2

Frayn M 2000 *Copenhagen* (New York: Anchor Books)

French A P and Kennedy P J (ed) 1985 *Niels Bohr: A Centenary Volume* (Cambridge, MA: Harvard University Press)

Heilbron J L and Kuhn T S 1969 *Hist. Stud. Phys. Sci.* **1** 211–90

Kragh H 2011 *Phys. Perspect.* **13** 4–35

Kragh H 2012 *Niels Bohr and the Quantum Atom: The Bohr Model of Atomic Structure 1913–1925* (Oxford: Oxford University Press)

Niels Bohr Archive 2002 Release of Documents relating to 1941 Bohr-Heisenberg meeting. https://nbarchive.dk/collections/bohr-heisenberg/

Pais A 1991 *Niels Bohr's Times, in Physics, Philosophy, and Polity* (Oxford: Oxford University Press)
Reed B C 2020 *Manhattan Project: Story of the Century* (Berlin: Springer)
Rhodes R 1986 *The Making of the Atomic Bomb* (New York: Simon and Schuster)
Schwarz W H E 2013 *Angew. Chem. Int. Ed.* **52** 12228–38
Sweet W 2002 *Bull Atom. Sci.* **58** 20–7

Appendix A

Bibliography

Aaserud F and Heilbron J L 2013 *Love, Literature, and the Quantum Atom* (Oxford: Oxford University Press)

Aaserud F and Kragh H (ed) 2015 *One Hundred Years of the Bohr Atom: Proc. from a Conf.* (Copenhagen: Royal Danish Academy of Sciences and Letters)

Ahmadi M *et al* 2017 Observation of the 1s–2s transition in trapped antihydrogen *Nature* **541** 506–10

Balmer J J 1885 *Ann. Phys. Lpz.* **261** 80–7

Beyer A *et al* 2017 *Science* **358** 79–85

Bernstein J 2003 *Phys. Perspect.* **5** 243–61

Bethe H 2000 *Phys. Today* **53** 34–6

Bohr N 1913a *Philos. Mag. Ser. 6* **26** 1–24

Bohr N 1913b *Philos. Mag. Ser. 6* **26** 476–502

Bohr N 1913c *Philos. Mag. Ser. 6* **26** 857–75

Bohr N 1913d *Nature* **92** 231–2

Bohr N 1936 *Nature* **137** 344–8

Bohr N and Wheeler J A 1939 *Phys. Rev.* **56** 426–50

Bohr N 1972 *Collected Works* (Amsterdam: Elsevier)

Brackett F S 1922 *Nature* **109** 209

Campbell G K, Leanhardt A E, Mun J, Boyd M, Streed E W, Ketterle W and Pritchard D E 2005 *Phys. Rev. Lett.* **94** 170403

Cassidy D C 2009 *Beyond Uncertainty: Heisenberg, Quantum Physics, and the Bomb* (New York: Bellevue Literary Press)

Cassidy D C 2017 *Farm Hall and the German Atomic Project of World War II: A Dramatic History* (Berlin: Springer)

Crawford F S 1989 *Am. J. Phys.* **57** 621–8

Davisson C and Germer L H 1927 *Phys. Rev.* **30** 707–40

Ellison F O and Hollingsworth C A 1976 *J. Chem. Educ.* **53** 767–70

Ford K 2018 *Phys. Teach.* **56** 196

Ford K 2018 *Phys. Teach.* **56** 500–2

Frayn M 2000 *Copenhagen* (New York: Anchor Books)

doi:10.1088/978-0-7503-3611-6ch6

French A P and Kennedy P J (ed) 1985 *Niels Bohr: A Centenary Volume* (Cambridge, MA: Harvard University Press)
Griffiths D J 1995 *Introduction to Quantum Mechanics* (Upper Saddle River, NJ: Prentice-Hall)
Heilbron J L and Kuhn T S 1969 *Hist. Stud. Phys. Sci.* **1** 211–90
Höglund B and Mezger P *Science* **150** 339–40 347–348
Humphreys C J 1953 *J. Res. Natl. Bur. Stand.* **50** 1–6
Kragh H 2011 *Phys. Perspect.* **13** 4–35
Kragh H 2012 *Niels Bohr and the Quantum Atom: The Bohr Model of Atomic Structure 1913–1925* (Oxford: Oxford University Press)
Kuhn T S 1987 *Black-Body Theory and the Quantum Discontinuity 1894–1912* (Chicago, IL: University of Chicago Press)
Lewis G 1926 *Nature* **118** 874–5
Lohmann B and Weingold E 1981 *Phys. Lett.* A **86** 139
Lyman T 1914 *Phys. Rev.* **3** 504
Mermin D 2004 *Phys. Today* **57** 10
Michelson A A and Morley E W 1887 *Philos. Mag. Ser. 5* **24** 463–6
Moore R 1985 *Niels Bohr: The Man, His Science, and the World They Changed* (Cambridge, MA: MIT Press)
Moseley H G J 1913 *Philos. Mag. Ser. 6* **26** 1024–34
Moseley H G J 1914 *Philos. Mag. Ser. 6* **27** 703–13
Nagaoka H 1904 *Philos. Mag. Ser. 6* **7** 445–55
Nicholson J W 1911 *Mon. Not. R. Astron. Soc.* **72** 49–64
Niels Bohr Archive 2002 Release of documents relating to 1941 Bohr–Heisenberg meeting. https://nbarchive.dk/collections/bohr-heisenberg/
Pais A 1986 *Inward Bound: Of Matter and Forces in the Physical World* (Oxford: Oxford University Press)
Pais A 1991 *Niels Bohr's Times, in Physics, Philosophy, and Polity* (Oxford: Oxford University Press)
Paschen F 1908 *Ann. Phys. Lpz.* **332** 537–70
Pfund A H 1924 *J. Opt. Sci. Am.* **9** 193–6
Pickering E C 1897 *Astrophys. J.* **5** 92–4
Reed B C 2005 *Phys. Educ.* **40** 117–8
Reed B C 2020 *Manhattan Project: Story of the Century* (Berlin: Springer)
Rhodes R 1986 *The Making of the Atomic Bomb* (New York: Simon and Schuster)
Ritz W 1908 *Astrophys. J.* **28** 237–43
Rozental S 1967 *Niels Bohr: His Life and Work as Seen by His Friends and Colleagues* (New York: John Wiley and Sons)
Rutherford E 1911 *Philos. Mag. Ser. 6* **21** 669–88
Rydberg J R 1890 *Philos. Mag. Ser. 5* **29B** 331–7
Schwarz W H E 2013 *Angew. Chem. Int. Ed.* **52** 12228
Soddy F 1913 *Nature* **92** 399–400
Sweet W 2002 *Bull. Atom. Sci.* **58** 20–7
Thomson G P 1928 *Proc. R. Soc.* A **117** 600–9
Urey H, Brickwedde F and Murphy G 1932 *Phys. Rev.* **39** 164–5

Appendix B

Fundamental physical constants (A = ampere, C = Coulomb, F = Farad, J = Joule, T = Tesla)

Name	Symbol	Value	Units	Equivalent units
Speed of light	c	$2.997\,924\,58 \times 10^{8}$	m s^{-1}	
Planck's constant	h	$6.626\,070\,15 \times 10^{-34}$	J s	kg m^{2} s^{-1}
Electron charge	e	$1.602\,176\,634 \times 10^{-19}$	C	
Permittivity constant	ε_o	$8.854\,187\,812\,8 \times 10^{-12}$	F m^{-1}	C^{2} s^{2} kg^{-1} m^{-3}
Electron mass	m_e	$9.109\,383\,701\,5 \times 10^{-31}$	kg	
Proton mass	m_p	$1.672\,621\,923\,7 \times 10^{-27}$	kg	
Neutron mass	m_n	$1.674\,927\,498 \times 10^{-27}$	kg	
Atomic mass unit	amu	$1.660\,539\,066\,6 \times 10^{-27}$	kg	
Boltzmann's constant	k	$1.380\,649 \times 10^{-23}$	J K^{-1}	kg m^{2} s^{-2} K^{-1}
Alpha particle mass	m_α	4.001 506 179 127	amu	
Avogadro's number	N_A	$6.022\,140\,76 \times 10^{23}$		
Permeability constant	μ_o	$1.256\,637\,062 \times 10^{-6}$	T m A^{-1}	kg m C^{-2} = N A^{-2}
Gravitational constant	G	$6.674\,30 \times 10^{-11}$	m^{3} $kg^{-1}s^{-2}$	

Source: National Institute of Standards and Technology Recommended Values of the Fundamental Physical Constants: 2018. https://physics.nist.gov/cuu/pdf/wall_2018.pdf.

doi:10.1088/978-0-7503-3611-6ch7

Derived constants and combinations of constants

Description/symbol	Expression	Value	Units	Comment
Factor in spiral-in time	$4\pi^2\varepsilon_o^2 m_e^2 c^3/e^4$	$1.050\ 16 \times 10^{20}$	m^{-3}	Section 1.4
$t_{collapse}\ t_{orbit}$ factor	$\pi^{1/2}\varepsilon_o^{3/2} m_e^{3/2} c^3/e^3$	$2.659\ 88 \times 10^{20}$	$m^{-3/2}$	Section 1.4
Bohr radius a_O	$\varepsilon_o h^2/\pi m_e e^2$	0.529 177 2	Å	Equation (2.6)
μ_H	$m_e m_p/(m_e + m_p)$	$9.104\ 425 \times 10^{-31}$	kg	H reduced mass, equation (1.11)
μ_H/m_e	$m_p/(m_e + m_p)$	0.999 455 7		
Hydrogen $n = 1$ energy	$\mu_H e^4/8\varepsilon_o^2 h^2$	$2.178\ 686 \times 10^{-18}$	J	Equation (2.8)
Hydrogen $n = 1$ energy	$\mu_H e^4/8\varepsilon_o^2 h^2$	13.598 287	eV	Equation (2.8)
Hartree (R_∞)	$m_e e^4/4\varepsilon_o^2 h^2$	27.196 574 53	eV	Section 2.3
R_H	$e^4\mu_H/8\varepsilon_o^2 h^3 c$	10 967 758.34	m^{-1}	Rydberg constant (H), equation (2.11)
$(R_H)^{-1}$ in Å	$8\varepsilon_o^2 h^3 c/e^4\mu_H$	911.763 342 1	Å	Inverse Rydberg constant (H)
$4(R_H)^{-1}$ in Å	$32\varepsilon_o^2 h^3 c/e^4\mu_H$	3647.053 369	Å	Balmer series limit
R_∞	$m_e e^4/8\varepsilon_o^2 h^3 c$	10 973 731.57	m^{-1}	Infinite-mass Rydberg constant
α	$e^2/2\varepsilon_o hc$	$7.297\ 353 \times 10^{-3}$	—	Fine-structure constant, section 3.2
μ_{He}	$m_e m_\alpha/(m_e + m_\alpha)$	$9.108\ 135\ 04 \times 10^{-31}$	kg	Helium reduced mass; section 2.4
Helium Rydberg constant	$e^4\mu_{He}/2\varepsilon_o^2 h^3 c$	43 888 909.39	m^{-1}	Equation (2.16)
Inverse He Rydberg in Å	$2\varepsilon_o^2 h^3 c/e^4\mu_{He}$	227.847 994 8	Å	Equation (2.17)
Orbital frequency factor	$m_e e^4/4\varepsilon_o^2 h^3$	$6.579\ 68 \times 10^{15}$	s^{-1}	Section 2.7
Deuterium (3–2) difference	$144\varepsilon_o^2 h^3 c/5m_p e^4$	1.786 65	Å	Section 4.1
λ_C	$h/2m_e c$	$2.426\ 31 \times 10^{-12}$	m	Compton wavelength; section 3.6
de Broglie wavelength	$2\varepsilon_o h^2/m_e e^2$	3.324 92	Å	Section 3.6
$n = 1$ magnetic field	$\pi\ \mu_o m_e^2 e^7/8\varepsilon_o^3 h^5$	12.516 824 42	T	Section 4.2
Bohr magneton μ_B	$eh/4\pi\ m_e$	$9.274\ 010\ 1 \times 10^{-24}$	$J\ T^{-1}$	Also A m^2; section 4.2

www.ingramcontent.com/pod-product-compliance
Ingram Content Group UK Ltd.
Pitfield, Milton Keynes, MK11 3LW, UK
UKHW051848100926
8556IPUK00005B/522

9 780750 336093